© 2006 Nike Zachmanoglou Tirman

About the Author

JOHN TIRMAN is the executive director of MIT's
Center for International Studies. He is the author,
or coauthor and editor, of nine books on international affairs. His work has appeared in the *New
York Times*, *Washington Post*, *The Nation*, *Wall Street
Journal*, and *International Herald Tribune*. He lives
in Cambridge, Massachusetts.

100 WAYS
AMERICA IS
SCREWING UP
THE WORLD

100 WAYS AMERICA IS SCREWING UP THE WORLD

JOHN TIRMAN

HARPER PERENNIAL

NEW YORK • LONDON • TORONTO • SYDNEY

HARPER ⬤ PERENNIAL

HarperCollins books may be purchased for educational, business, or sales promotional use. For information please write: Special Markets Department, HarperCollins Publishers, 10 East 53rd Street, New York, NY 10022.

FIRST EDITION

Designed by Nancy Singer Olaguera

Library of Congress Cataloging-in-Publication Data is available upon request.

ISBN-10: 0-06-113301-9
ISBN-13: 978-0-06-113301-5

06 07 08 09 10   ISPN/RRD 10 9 8 7 6 5 4 3 2

For Coco

Contents

Foreword

Our country has soaked itself for a long time in the warm bath of self-congratulation. It started very early, as John Winthrop, the first governor of the Massachusetts Bay Colony, said in the year 163?: "We shall be as a city upon a hill—the eyes of all people are upon us."

The arrogance of that statement was demonstrated almost immediately, because six years later the soldiers of the Bay Colony began burning the villages of the Pequot Indians. William Bradford, a contemporary, described the massacre of one village's inhabitants:

> Those that escaped the fire, were slain with a sword; some hewed to pieces, some run through with spears, as they were quickly dispatched and very few escaped. It is conceived that thus were destroyed about 400 at this time. It was a fearful sight to see them frying in the fire and the streams of blood quenching the same but the victory seemed a sweet sacrifice and I gave prayers thereof to God, who had fought so wondrously for them and given them a speedy victory over so proud and insulting an enemy.

The idea that we make war "with God on our side" (in Bob Dylan's haunting words) runs through our history. On the eve of the Mexican war of 1846, an American journalist declared that the annexation of Texas was "the fulfillment of our manifest destiny to overspread the continent allotted by Providence for the free development of our yearly multiplying millions."

President William McKinley in 1898 told a group of visiting ministers that he had decided to take over the Philippines after he got down on his knees and prayed to God, and God told him it was the duty of the United States "to civilize and Christianize" the Filipino people. It is estimated that at least half a million Filipinos died as a result of our civilizing mission.

But reality has never brought any hesitation in the official rhetoric of self-praise. At the very time that the United States Army was carrying out massacres in the Philippines, Secretary of War Elihu Root was saying: "The American soldier is different from all other soldiers of all other countries since the world began. He is the advance guard of liberty and justice, of law and order, and of peace and happiness."

It becomes easier to admire yourself if you change the facts of history. President George W. Bush, invited a century after the Philippines war to speak to the Philippine National Congress, told his audience: "America is proud of its part in the great story of the Filipino people. Together our soldiers liberated the Philippines from colonial rule."

Woodrow Wilson is represented in most of our history books and our texts on international affairs as standing for "idealism," but such a characterization must surely be reconsidered when one looks at Wilson's actual behavior: his bombardment of the Mexican city of Vera Cruz in 1914; his military occupation of Haiti in 1915, killing thousands of Haitians who resisted; his occupation of the Dominican Republic in 1916; the entrance of the United States into the slaughter fields of the war in Europe in World War I.

All of this was surrounded with the language of righteousness. Indeed, it was just a few months after the bullying attack on Mexico in 1914 that Wilson told graduating midshipmen at the Naval Academy that the United States used "her navy and her army . . . as the instruments of civilization, not as the instruments of aggression." The idea of America, Wilson said, "is to serve humanity, and every time you let the Stars and Stripes free to the wind you ought to realize that is in itself a message that you are on an errand which other navies have sometimes forgotten; not an errand of conquest, but an errand of service."

The victory in World War II gave a special boost to the idea that the United States had a special dispensation to press its power on the world. Now, said Henry Luce, the powerful owner of *Time, Life,* and *Fortune,* the United States had "the right to exert upon the world the full impact of our influence, for such purposes as we see fit and by such means as we see fit." The twentieth century would be, Luce declared, "The American Century."

Indeed, the United States, through the second half of the twentieth century, set out to fulfill this boast. Its corporations roamed the world. Its armed forces fought a ten-year war in Vietnam, and short wars in Panama, Grenada, and Iraq. It engaged in covert actions, to overthrow democratically elected governments in Iran, Guatemala, Chile. And it supported military dictatorships in Iran, the Philippines, Indonesia, and the Caribbean.

Though it failed in its attempt to overthrow the Cuban government of Fidel Castro, and was forced to withdraw from Vietnam, the United States continued to maintain hundreds of military bases throughout the world. Surrounding its aggressions with rhetoric about terrorism and democracy, it set out to establish more bases in the oil-rich Middle East.

In 1991, with the Soviet Union collapsing and no longer a rival military power, the elder Bush, in his State of the Union address in 1991, said there would be a "new world order." He called upon his own version of history to say: "For two centuries America has served the world as an inspiring example of freedom and democracy."

Democratic presidents have been just as extravagant in their praise for United States behavior in the world, and just as confident that it is the mission of the United States to export its virtues to other countries. Bill Clinton, speaking at the West Point commencement in the spring of 1993, said: "The values you learned here . . . will be able to spread throughout the country and throughout the world."

This idea was reiterated by George W. Bush in his inaugural address in January 2005: spreading liberty around the world he said is "the calling of our time." The *New York Times* said his speech was "striking for its idealism." The day before Bush uttered those words, there was a photo in the *Times* showing a small Iraqi girl crouching, covered with blood, screaming. Her parents had just been killed when U.S. soldiers fired on their car when it failed to stop.

One might expect our political leaders to sing the praises of this country. Self-praise comes with the job. But the idea that the United States is an especially virtuous country has also been expressed by many American intellectuals. Even as they recognize that the United States is an imperial power, asserting its dominance over the rest of the world, they find this to be different than the imperialism of the British Empire, the French and Dutch and German empires.

The political commentator Charles Krauthammer has written: "We are a unique benign imperialism." Michael Ignatieff, of Harvard's Kennedy School, said: "The twenty-first century is a new invention . . . a global hegemony whose grace notes are free markets, human rights, and democracy." These statements were made even as the United States was carrying on a war in Iraq that, after three years, and with no end in sight, had led to anywhere from 30,000 to 100,000 Iraqi deaths.

The "city on a hill" proclaimed by John Winthrop long ago indeed has had, as he said, "the eyes of all people" on us. But despite the obvious accomplishments of this country—its advances in technology and science, its impressive standard of living for much of its population, its relative freedom of expression—those eyes have been less and less admiring.

The United States has been increasingly viewed by people all over the world as a rogue nation—a government increasingly contemptuous of international agreements, insisting, as laid down in the National Security Strategy of September 2002, that it has the unilateral right to use force to eliminate any potential threat to U.S. dominance, and "will, if necessary, act preemptively."

Indeed, this is what was done in March of 2003, when the United States bombed and invaded Iraq, without the approval of the United Nations, and, indeed, in violation of the U.N. charter, which permits the use of violence only in self-defense. Again distancing itself from the community of nations, the United States has refused to sign the treaty banning land mines, agreed to by over a hundred other countries. It has refused to cooperate with the International Court of Justice, and in 1995, in Clinton's administration, refused to strengthen the Biological Weapons Convention, adopted by 145 nations.

Alone among the advanced industrial nations, it has refused to sign the Kyoto Protocol to try to slow the process of global warming. As Donald Kennedy, editor of the respected publication *Science* has said, the United States has taken "a contemptuous pass on multilateral engagement with the global warming problem." It has formulated a "Strategic Master Plan to put weapons in space, in violation of international treaties for the neutralization of space."

When, in 2005, the State Department issued its annual report on abuses of human rights, naming countries that violated such rights, its spokesperson said: "promoting human rights . . . is the bedrock of our policy and our foremost concern." But there were indignant responses from around the world, pointing out that the United States was missing from the list, despite clear evidence of the use of torture against "suspects." A Turkish newspaper commented: "There's not even mention of the incidents in Abu Ghraib prison, no mention of Guantanamo." A newspaper in Sydney, Australia, pointed out that the United States sends suspects—people who have not been tried or found guilty of anything—to prisons in Morocco, Egypt, Libya, Uzbekistan, countries that the State Department itself says use torture.

In the foregoing few pages, and in John Tirman's book, the strong criticism of our nation's policies may be troubling to those readers who, like Tirman and myself, and virtually all Americans, grew up with the idea that our country is uniquely virtuous. In first grade we pledged allegiance to the flag and the country, which promised "liberty and justice for all." And we sang "The Star-Spangled Banner," proclaiming that this is "the land of the free and the home of the brave," surely not thinking that those words were written at a time when the nation included several million slaves.

Some readers may conclude that there is something unpatriotic about pointing to slavery, racism, economic injustice, the massacre of Indians, the imperial wars. I welcome John Tirman's book because I think we have for too long congratulated ourselves for our good deeds (and there have been some, as Tirman points out) and ignored the ways we have violated human rights throughout our history up to this very moment. I believe we do our country a service when we look at it honestly, free of nationalist arrogance.

We need to think carefully about what it means to be patriotic. If patriotism means uncritical support of what your government does, and if it is unpatriotic to criticize the government, then patriotism fits in perfectly with totalitarianism. But if we are to live in a democracy, we should recall the principles of the Declaration of Independence, which declared that governments are artificial entities, set up by the people of the country to achieve certain ends—an equal right of all to life, liberty, and the pursuit of happiness.

And, the Declaration says, "Whenever any form of government becomes destructive of these ends, it is the right of the people to alter or to abolish it." If we have the right to "alter or abolish," then we certainly have the right to criticize. This formulation of democracy distinguishes between the government and the country, and suggests that true patriotism requires allegiance to the principles for which the country is supposed to stand.

There have always been patriotic Americans who understood this, who refused to consider this country as deserving of special praise, who have insisted that the United States has been given no special dispensation to expand its power at the expense of other people, who have resisted the temptations of national arrogance.

The ex-slave Frederick Douglass, speaking in 1852 on the Fourth of July, asked a white audience in Rochester, New York: "What, to the American slave, is your Fourth of July? I answer: a day that reveals to him, more than all other days in the year, the gross injustice and cruelty to which he is the constant victim. To him your celebration is a sham; your boasted liberty, an unholy license; your national greatness, swelling vanity . . . your shouts of liberty and equality, hollow mockery. . . . There is not a nation on the earth guilty of practices more shocking and bloody than are the people of the United States, at this very hour."

Douglass's fellow abolitionist William Lloyd Garrison also refused to accept the idea of a special allegiance to the nation. He wrote: "My country is the world. My countrymen are mankind." Henry David Thoreau, provoked by the war in Mexico and the nationalist fervor it produced, wrote: "Nations! What are nations? . . . Like insects, they swarm." In our time, Kurt Vonnegut (in his novel *Cat's Cradle*) writes of unnatural abstractions he calls "granfalloons," among which he includes "any nation, any time, anywhere."

Martin Luther King Jr., whose birthday we celebrate every year throughout the nation, recognizing him as one of the great prophets of our country, did not hesitate to criticize his government in the harshest of terms. As the war in Vietnam raged, King spoke of "the greatest purveyor of violence in the world today—my own government."

It is in that admirable tradition of dissent, defiance, resistance to illegitimate authority, insistence that not only Americans, but people everywhere have a right to life, liberty, and the pursuit of happiness, that John Tirman has written this book. To write it, read it, publish it, is a celebration of democracy.

—Howard Zinn

Introduction

During times of universal deceit, telling the truth becomes a revolutionary act.

—*George Orwell*

Has America really screwed up the world in one hundred ways, more or less? It might seem cranky or clownish (or both) to describe America's role in the world in such disrespectful terms, but there's a serious point to this: like any enormously powerful nation, the United States has thrown its weight around in countless ways; its culture and products increasingly dominate culture and consumer choices around the world; its narrow economic agenda is often force fed to the multitudes of institutions that regulate global commerce, and, through all this, it congratulates itself as the sole beacon of freedom and goodness in the history of humanity. The American ideals of liberty, equal opportunity, political rights, and the pursuit of happiness have anchored many dreams around the world for two centuries, and I bow to these monumental achievements. But we know that story, because it is intoned like a background chant in the daily life of the nation. What we don't know so well is how and why so much of what America does in the world has gone awry, or worse, has gone as planned with disastrous results for almost everyone else.

So look at this little book not merely as contrarian but as corrective. We need a dose of reality every now and then to counterbalance the saccharine self-love, and we need to think anew about how America interacts with the rest of the planet. Every school-child knows that we live in a "globalized" world, made smaller and more connected by technology, migration, and trade. What we do not fully appreciate, however, is that the globalization is shaped by major powers and the United States particularly, and is shaped for the benefit of a small number of Americans. What, exactly, is wrought by the tendrils of globalization? Who wins, who loses? And because America is simply big, rich, and dynamic, much of what we do has worldwide impact, neither malevolent nor intentional, but still resonating every day in many corners of the globe.

It scarcely needs to be said, but I'll say it to fend off the predictable yelps of wounded pride, that America is not the only one making the world less safe, fair, diverse, or clean. There are plenty of culprits, some old—the British Empire, the Soviet Union, religions generally—and some new, such as al Qaeda. But America is my country—born, bred, and corn-fed—and as the pundits like to say, over and over again, the United States is the world's only superpower. I hold it to a higher standard.

There will be others who say this is an anti-American rant, the whine of the left, and other potshots. Consider for a moment another possibility. Think of this as simply truth telling inside a family that needs to hear it, tough love for the twenty-first century.

A few acknowledgments are in order. To John Williams and colleagues at HarperCollins, for their thoughtful support; to Howard Zinn, for his foreword and his exemplary life; to Seamus McKiernan, who provided research assistance; to many friends and colleagues for coming up with ideas; and to my family, for tolerating a long stretch of weekends and early mornings when I disappeared into the attic to write.

So let's see what hundred or so ways this actually plays out. Some of these are fun, but still provocative, and most of them are deadly serious, even if playfully explained. I think of this book as starting a conversation, one that readers can join by entering my Web site, www.johntirman.com, where I also have references for the *100 Ways* and other features, including links to organizations trying to solve problems identified in this book.

Most of all, I hope that my fellow citizens will find these *100 Ways* informative and thought provoking, a long overdue self-examination, which will stimulate discussion and action in the years to come.

100 WAYS
AMERICA IS
SCREWING UP
THE WORLD

1 Altering the Earth's Climate

Some acts of a powerful nation affect people somewhere in the world very directly, like starting a war. Sometimes a lack of action has a deplorable effect, such as not stopping genocide. Slowly unfolding and irreversible impacts, often unintentional, are also devastating, as with the gradual loss of cultural diversity. And at times there's a head-in-the-sand ignorance or neglect of a visible problem, such as the first decade or so of responses to the HIV/AIDS epidemic.

Rarely do all four of these phenomena combine noticeably into one. But with climate change, the United States has managed almost single-handedly to be the cause, the obstacle to remedial action, a chronic ignoramus, and an aggressive denier of its monumental culpability.

No other issue on the global table today affects the well-being of everything on the planet as much as the dumping of greenhouse gases into the atmosphere, and, along with it, into the oceans. It will cause incalculable human suffering and economic costs. It will touch everyone and everything except perhaps the very rich, but even they will likely be adversely affected. It is preventable—or at least can be mitigated—but we ignore it. Understanding why speaks volumes about how America acts in the world today.

Let's begin with a few facts. The relevant changes in the earth's climate, including the warming of the atmosphere, are caused by emissions of carbon dioxide from industrial plants, automobiles, and other technologies created by humans. As the earth warms, the ice caps melt and the seas rise. The weather is likely to change—not only warmer, but more volatile, resulting in more droughts and forest fires. Higher levels of carbon dioxide in the air also can affect crops, livestock, and the transmission of diseases. Carbon dioxide in the air is absorbed by the oceans, as is freshwater runoff from melting, and this (which has been carefully measured already) is likely to alter ocean temperatures, currents, and the viability of marine life. The oceans, in fact, which control much of what happens climatically above the waterline, may be the key: when ocean current circulation was disrupted by glacier melting 12,000 years ago, it ushered in the last Ice Age.

The consequences of these changes are difficult to know, not only because the pace of change is unpredictable, but because of the scale of the systems. Ecosystems are dynamic: they interact with one another in millions of measurable and immeasurable ways. For example, as the permafrost melts, it not only adds to sea levels, but reduces the amount of sunlight reflected back into outer space, sunlight instead being absorbed into the waters to produce more warming in a continuous feedback loop. A similar dynamic is visible in microbes in the soil: longer growing seasons due to warming enable them to produce and release more methane into the atmosphere, which altogether is an amazing quantity.

Because the systems are so large, the visible impacts are seen only gradually; in fact, there is a "thermal inertia" at work, meaning that the effects will manifest gradually and continue for centuries, even if we halted the growth of greenhouse gases immediately.

Within that conservative range of estimates, however, the effects are certain to be enormous. The loss of biological diversity, the costly transformations in agriculture, the potential extinction of 10 percent or more of all species, the wholesale adjustments of living and working in the coastal cities, the freezing effects on northern Europe resulting from the loss of the gulf stream, the growth in new virulent diseases due to rising temperatures—all of these effects are now considered to be probable, not merely imaginable, over the coming decades.

The science is definitive. There is no major dispute among the climatologists, oceanographers, and others studying climate change, a field of work that is now well developed. The scientific deniers are skeptics about the potential scale of destruction, not the fact that it's occurring.

The main culprit in all this is the industrial age itself, the use of fossil fuels in particular—coal and petroleum—to run the factories, cars, and electric power plants of the world. Since the entire world has been industrializing for more than two centuries, one could say the whole of humanity is to blame. But the United States holds a special place in this pantheon of pollution.

America is the largest polluter in the world, and no one really comes in a close second. We produce more greenhouse gases—the carbon dioxide from burning fossil fuels that produce the "greenhouse effect" of global warming—than any other country. Our 4 percent of the world's population produces 25 percent of all carbon dioxide emissions. We produce as much as the rest of the industrial world combined, several times more per person than Britain or Japan. And the United States will continue to be, on current trends, the largest contributor to the problem for many years to come.

To their credit, the United Nations and most countries in the world have made attempts to deal with this looming catastrophe. The major effort of this kind was the Kyoto Protocol, which was a modest constraint, asking for nations to reduce, by 2012, greenhouse emissions to slightly below 1990 levels. The treaty was finalized in 1997 and by the end of 2005 had been signed and ratified by 157 countries, including all of the European Community, Russia, Canada, and Japan. That is every country in the G-8, the largest industrial countries, except one: the United States of America. Why?

Given how widely supported the issue is among Americans—clear majorities support the Kyoto treaty and more stringent limits on carbon emissions—the absence of action by President Bush (and the weak leadership shown by President Clinton) is puzzling. But strong economic forces are arrayed against action, and since 9/11 the public has been distracted by the threat of terrorism, a problem that is minuscule compared with climate change.

The strategy of the deniers, with corporate backers, has been to cast doubt on the scientific consensus that has taken shape and to overstate the short-term costs of immediate action. It has created phony front organizations that issue "studies" and generate news that is reported on equal footing with the major scientific institutions. By sowing doubt among the public about the scientific consensus and the immediate urgency of the problem, the corporate lobby has diminished the saliency of the issue.

President Bush has reflected this strategy, among others. In a letter in 2001, he said that "I oppose the Kyoto protocol because it would cause serious harm to the economy." Just prior to the G-8 summit in 2005, he said that "the Kyoto treaty would have wrecked our economy," while admitting that human activity was "to some extent" to blame for climate change. He has proposed to reduce by 18 percent the *increase* in American contributions to greenhouse gases, which themselves will have gone up by one-third between 1990 and 2012. And the United States will not budge before China and India are included, shifting the blame for emissions to these two populous and rapidly industrializing countries. They will become problems eventually, but America remains the most carbon-intensive country for the near term. And getting China and India to invest now to reduce greenhouse emissions will only be achieved if the United States takes the lead.

Bush says that the development of new technologies is the key to dealing with the problem, an old "technical fix" mentality that would have to include a vast expansion of doubtful technologies like nuclear power plants. When other countries improve efficiency and reduce emissions, they will leap ahead on a range of performance measures that will greatly enhance their productivity and their economies, while the United States remains wedded to petroleum-based fuels that are both insecure and polluting. We can see how serious Congress and successive presidents have been about technical fixes by their attitude toward automobile fuel efficiency: it could be increased threefold overnight but is blocked by the friends of the auto and oil industries, including Mr. Bush.

At least he hasn't said yet, as Ronald Reagan did, that "trees cause more pollution than automobiles do." What is just as interesting as the denial of scientific consensus on the issue is the role reversal of "conservatives." The putative philosophy of conservatism is, as the name implies, to conserve society and its institutions, to oppose radical change, to be wary of human actions that will upset the heritage we have been given. Nothing will more powerfully alter society and institutions and the natural heritage human civilizations have been bequeathed than climate change. To *not* act to conserve is what is reckless. Taking responsibility for our actions would also seem to fit into this mix. But this philosophy of profits above all else is what now distinguishes today's American right wing from the conservative philosophies of the eighteenth and nineteenth centuries that are supposedly their forebears.

It would cost little to act prudently to reduce greenhouse gases. New products and services would be created, and along with them, new "green" jobs. Where there are more costly measures to be taken, we keep in mind that they are discounted compared with dealing with problems later on. This is already a well-established fact of environmental economics. So investments in efficiency would be good for all kinds of reasons, as they almost always are. And this would be the "conservative" approach.

Once global warming and climate change became known nearly twenty years ago, we should have acted. We were producing the most greenhouse emissions. As the world's industrial giant, as the defender of human rights, as the avatar of science and technology, America would have been the logical leader. As the Cold War ended, it would have been the perfect cause for a new American globalism: challenging ourselves and others to create a sustainable and equitable economy worldwide.

But no. We've gone backward, and continue to slide downhill, and our irresponsibility affects everyone on the planet. The costs, dangers, and suffering will far surpass any other problem before us today.

2   Television

In the last century, few inventions have altered the way we live our daily lives more than television, and no country has had more to do with that than America. I recall traveling through Mozambique to its capital, Maputo, about ten years ago, the country still ravaged from its long civil war, with burned out buildings and ruins marring every block. We arrived at our modest hotel, sat down to rest, and turned on the small television set in our room, and, presto, there were Jerry, George, Elaine, and Kramer. It seemed emblematic of so many things, not least of them the cultural globalization—some would call it hegemony—that American television, even more than Hollywood or the Internet, represents.

One billion television sets have been sold worldwide since the first ones appeared in the 1920s. The growth in the last three decades of the twentieth century was phenomenal. In China, now the biggest couch potato country, "set ownership increased from a few thousand to 260 million between 1960 and 1994. And from 1975 to 1994 viewers increased from 18 million to 900 million," says one report. Of all American households, something like 98 percent are plugged in. Television is on, it seems, everywhere and all the time.

While there is much to lament about television programming, there is also much to celebrate. It is a given now that national and even global moments of importance are shared via television. I can recall with total clarity the reports of John F. Kennedy's assassination and his funeral, the first moon landing, and the towers falling on 9/11. For devotees of sports and entertainment, television is a godsend. It has transformed our ability to see and feel and hear the rest of our country and the world.

But it is far from benign, or neutral, or even a net gain. It's not just that television programs are often mind-numbing for the masses, filled with useless, boring nonsense or worse. After all, many pastimes are stupid and wasteful. Television is actually more pernicious than its renowned reputation for delivering junk food to the brain.

More than anything in society—global, national, and local societies—television separates us from one another. It isolates us, not only from our neighbors and communities, but from our families and friends. It creates a solitary world of viewing, a passive state of receiving images and opinions and a sprinkling of information. This receiving is done essentially alone, and that is socially and politically lethal.

People tend to socialize less with their neighbors, a slowly unfolding but long-term trend, and within their communities. This kind of socializing is called social capital—"the complex network of human interactions and community connections that lead to mutual support, cooperation, trust, institutional effectiveness." It is very important to the healthiness of a community, a town or city, and a country. People learn about their communities, become emotionally and socially invested in them, through personal interaction. They form what was once commonly referred to as solidarity. We learn from one another, like one another, build common cause. A good summary of the concept is provided by Eric Uslander, a political theorist at the University of Maryland:

> Communities with strong positive values (including trust in others) and ties that bind people to one another will have more powerful norms of generalized reciprocity and cooperation. Trust as a moral resource leads us to look beyond our own kind. It means that we downplay bad experiences and cooperate even when we are not sure that others will oblige. Trust makes for a vibrant community in several ways. Trust promotes cooperation. It leads people to take active roles in their community, to behave morally, and to compromise. People who trust others aren't quite so ready to dismiss ideas they disagree with. When they can't get what they want, they are willing to listen to the other side. Communities with civic activism and moral behavior, where people give others their due, are more prosperous.

I quote this at length because it is crucial to understanding many social and political issues today, and it is neither a left nor right issue. Both sides embrace the nourishing aspects of social connectedness.

But social capital is in decline. It is declining, in part, because of the "technological transformation of leisure," where television and DVDs and video games and such substitute for having fun with other people. (Even work is transformed by many of these same technologies, with the Internet being a variation of TV.) Membership in civic organizations, PTAs, many kinds of sports leagues, and the like have declined since the 1960s. According to a Roper Report study, the number of Americans who report that "in the past year" they have "attended a public meeting on town or school affairs" fell by more than a third between 1973 and 1993.

Similar (or even greater) relative declines are evident in responses to questions about attending a political rally or speech, serving on a committee of some local organization, and working for a political party. Similar reductions have taken place in the numbers of volunteers for mainline civic organizations such as the Boy Scouts (off 60 percent since 1970) and the Red Cross (off 61 percent since 1970).

Television is a primary cause.

We are in the second generation of people who grew up with television, and it is now having life cycle effects in which this kind of separation and isolation has been ingrained and nurtured from an early age. We all see and decry the statistics of children watching five or six hours of TV every day, the tube as babysitter, the effect of the technology and imagery itself on cognitive development. It also, with the Internet, divorces children from socializing, from developing social skills and kinships that are so crucial to them as adults, too.

Television not only robs us of enriching and satisfying friendships and a sense of community, but it drains away the capacity and willingness to act politically, to be citizens in the fullest sense of the term.

In the last three decades, there has been a decline in direct engagement in politics (as evidenced by voter turnout and other indicators). Robert Putnam, the Harvard professor who documented so much about social capital and its importance, argues that modern mass media tend to exhaust the public of political energy, and allow elected officials to be less accountable. A kind of media malaise effect takes hold, fostering disillusionment, political apathy, distrust, and alienation. General voter trust in government institutions (including the military) has risen in the same period.

This is not about a liberal media bias (which is nonexistent) or permissiveness in Hollywood. It has to do with the nature of television viewing itself, how it affects our daily lives.

Heavy TV watching leads viewers (even among high educational/high income groups) to have more homogeneous or convergent opinions than light viewers, who tend to have more heterogeneous or divergent opinions. The "cultivation effect" of television viewing is one of making political opinions more uniform, and less tolerant of innovation. Media-cultivated facts and values have become standards by which Americans judge personal experience, family, and community behavior.

That's not to say the content is unimportant. People who are isolated from others tend to believe what they see on television more readily, and tend to be more easily manipulated. Because so much of what passes for news and information is paranoid, sensationally violent, superficial, and ethnocentric, individual viewers will tend to follow suit. Viewers meld the entertainment programs with real-life programs; and because entertainment programming dwells on cop shows, this becomes a reality in itself. But news programs themselves tend to obsess about crime, "damsels in distress," threats of terrorism, and similar frights that have little connection to an average person's daily life. These emphases increase the alarm and concern about these remote phenomena among the general public. Heavy viewers tend to see the world as a more dangerous place than those who watch TV less frequently. And television reinforces this, because without social interactions that can check such distortions of real life, TV reality becomes ever more dominant.

This has obvious dangers for a democracy, for America's place in the world, and for the world itself. The decline of social capital is reported in Europe, too, and while the correlation to television is not as strong, it is a factor. But the dynamic of television is so strong, and the forces of social bonding so vulnerable (think of repression's effects in China, for example), that television is now poised as the stronger force.

While we can't lay all this at America's feet, its pioneering role in television is definitive. That is, it has defined how television programming and habits have unfolded. Television programs in Europe look remarkably like America's. Developing countries are fed American programming above all else, especially in Anglophone countries. Corporate concentration in the news media guarantees that certain kinds of programming—mindless entertainment particularly—will dominate, will be exported (it is highly profitable, since production costs have already been covered), and will come to shape the viewing habits of increasing billions of people. CNN and other global networks shape news coverage, or replace it.

Television is one of those forces that seem inexorable and beyond repair. It is neither, of course, but the pathways to correcting its inimical consequences are hard to find. Nearly all lead through and from America, however; and here, a cultural earthquake would be needed to dislodge its grip.

3   The Cold War

When Harry Truman and Joseph Stalin kicked off the Cold War in the late 1940s, they probably didn't realize what a long game it would be—fully forty years of foot stomping, captive nations, petty dictators, surrogate wars, massive human-rights violations, wasted resources, and, of course, the nuclear arms race. The Cold War stands as the destructive centerpiece of the twentieth century, second only to Hitler's and Tojo's murderous lunacy. It was a stage for easy posturing while the drama itself wrought immense havoc. And its force field of waste, militarism, and fraudulent ideas continues to beset the world.

Containing Stalin's expansionist tendencies was a sound idea, but the political, diplomatic, and economic encirclement urged by the architect of the policy, George Kennan, soon became a military doctrine and, inevitably, a military circus. Republican president Dwight D. Eisenhower, no wilting flower when it came to the armed forces, warned at the end of his presidency of the "undue influence of the military-industrial complex." More prescient words were never spoken by any occupant of the White House.

The United States pursued a military strategy that proved unprincipled, counterproductive, and financially ruinous. It emphasized not only nuclear deterrence, but arming vicious strongmen the world over. This resulted in genocides in Guatemala and Cambodia, massively lethal wars in Angola and Mozambique, whole regions roiled by repression and conflict. "Triumphs" like supporting the Afghan mujahideen or the coup to install the shah of Iran turned against us. Direct wars in Korea and Vietnam took huge tolls—one million dead in Vietnam, a Korea still divided. The financial costs of the Cold War have been conservatively calculated in the trillions of dollars, and opportunity costs for American education, cities, infrastructure, and the other crumbling elements of our country are glaring. The costs in political liberty, in dashed democratic hopes in Latin America, much of Africa, Iran and Pakistan, Southeast Asia, and elsewhere cannot be calculated.

This is not an argument of moral equivalence. The Soviets were just as nasty abroad, and nastier at home. But the point is not to say who triumphed or who was worse or who spurred this or that phase of the hostility. Rather, it was almost all unnecessary, especially after Stalin died in 1953 and reformers came to power later that decade in Moscow. Despite that, the serious costs of the Cold War—the nuclear arms race, which only got going in the 1960s, the proxy wars in Africa, Indochina, and Latin America, and the grip on various surrogate strongmen—were accrued long after Stalin was gone.

Was there another way? As it happened, the ideals of American democracy and prosperity triumphed in spite of the military waste and belligerency. The Soviet system was always most vulnerable to ideas rather than missiles. The growing disparity between Western lifestyles and the deprivations in the USSR, including political and cultural freedoms, became increasingly apparent. Civil society activists in Western Europe—greens, peaceniks, feminists, etc.—cultivated their counterparts in Eastern Europe with a simple, compelling logic: the Cold War is bad for everybody, so let's demand its end. Europe was the buffer for the East-West rivalry, overflowing with troop deployments and spackled with nuclear weapons, and Europeans themselves were tired of the two behemoths threatening each other across their fences. By the late 1970s, they had had enough, and were saying so in large numbers and with great clarity.

American politics would not listen, of course, being in the habit of embracing Europe and remaining deaf to it at the same time. Détente did not sit well with much of the American policy and opinion elite, especially as U.S. misadventures and strongmen in Vietnam and the rest of Southeast Asia, the Persian Gulf, Central America, and parts of Africa began to unravel. Jimmy Carter, stunned by the Islamic revolution and the consequent Soviet occupation of Afghanistan, stoked up the arms race and ended détente, policies confirmed and strengthened by his successor, Ronald Reagan.

But the toughening at the top merely stirred the activism from below, and the peace movements in Europe and America, the growing boldness of civil society activists in Poland, Czechoslovakia, East Germany, and Hungary, and a broad, global opprobrium regarding the arms race began to have an effect. The new human rights movement achieved an enormous breakthrough with the Helsinki Accords of 1975 (opposed by Reagan and halfheartedly endorsed by Democrats), which proved to be a powerful device to carve out space for activism in the Soviet bloc and build strong global norms more widely. All the while, transnational activism—among scientists, for instance, on both sides who were worried by the nuclear brinkmanship—created its own opportunities, convinced opinion elites that the governments were out of step, and paved the way for broad public support for an end to the hostilities. When Mikhail Gorbachev came to power in 1985, he was seeking reform, but instead witnessed (and, importantly, allowed) a revolution from below. It finally brought down the rotting Soviet Empire. Reagan, chastened by the Iran-contra scandal, saved his presidency by embracing Gorbachev's reformism and witnessed the activists lead the way to the end of the Cold War. With some luck, it all could have happened twenty-five years earlier.

The misapprehension of the Cold War as a military victory for the United States has already led to new and tragic mistakes. The Cold War incurred colossal costs, too great, really, to calculate. They remain with us today—nuclear weapons, overbuilt militaries, rogue intelligence agencies, human rights violations, failed states, and cultures of violence and reaction throughout much of the world.

America's role in the Cold War is not one to glorify—if it was occasionally necessary, in some way, this need was always overpursued and its noxious practices made permanent. No, rather, the Cold War should be a phenomenon to learn from, a pedagogical opportunity we have yet to engage.

4   Dumping Toxins

Evil comes in many packages. It can pilot airplanes into skyscrapers, commit genocide against innocent peoples, and produce mass delusions of righteousness. But sometimes evil appears in simple and stark forms. Over decades now, American corporations have been exporting chemicals and other products proven to be hazardous or poisonous, to people in the third world. Congress and presidents have done precious little to stop this practice of "dumping." Millions of people the world over have suffered needlessly as a result.

Consider a product that everyone can agree is poisonous: pesticides. At least since the 1970s, activists have warned of these hazards—chemical companies selling overseas the pesticides that are banned in the United States. Some attempts to constrain this practice have been set back by legislative loopholes or White House artifice. In 1981, President Carter signed an executive order restricting the export of substances banned in the United States, an order rescinded by President Reagan when he was inaugurated a few days later.

Most U.S. efforts to control pesticide use in the third world were driven by concerns that agricultural products laden with pesticides were being imported into American markets.

The industry produces $30 billion worth of poisons each year, with hundreds of thousands of deaths associated with unintended exposure. About one in seven agricultural workers reports acute pesticide poisoning each year in the developing world, and 99 percent of the deaths from pesticide exposure occur in the global south. Children are particularly at risk, and malnourished children—200 million in the developing world—are the most vulnerable of all. Cancers and other deadly or chronic diseases, such as immune deficiencies, can result from exposure. Direct contact comes from spraying or stockpiling, and indirect contact from ingestion or groundwater contamination.

The chemical industry and its advocates have long maintained that the pesticides are a net benefit, having controlled pests and increased food production. But pests become resistant, and alternative farming methods are proving that the large doses of poisons the industry exports are unnecessary.

The pesticide problem parallels another, broader dumping issue, that of toxic industrial wastes that are deposited in the third world. The depth of the problem was revealed in several books and media reports in the 1980s, and the international community did take action, passing four treaties to control or prevent toxic waste dumping. The Basel Convention, which entered force in 1992 and has since been updated, is considered to be the most important of these; it bans all hazardous waste exports from developed to developing countries, and, as a result, the volume of such dumping has been dramatically reduced. There are loopholes aplenty, however, which include dumping of millions of components of computers and cell phones, for example, which contain many toxic chemicals.

Most of the countries have ratified the treaty; the United States is one of the few that has not. Circumventing the Basel Ban is a routine stratagem of exporters. A second key treaty, banning the "dirty dozen" of hazardous materials (including PCBs and dioxins), known as persistent organic pollutants, was signed and ratified by ninety-eight countries and went into force in 2004. The United States has yet to ratify this accord, either; President Bush made a great show of endorsing the treaty in 2001 but has not pressed for Senate action. The World Wildlife Fund has called for an additional twenty such pollutants to be added to the banned list.

The scale of the destruction and human toll these kinds of chemicals and pesticides has exacted is almost impossible to estimate. As the activist Basel Action Network notes, "The mass migration of the 'effluent of the affluent' in the name of development, globalization, and free trade is in fact a violation of environmental justice and can be considered a crime against the environment and human rights." Perhaps even a crime against humanity, to use a commonly understood legal term.

5   Market Mantra

The Tragic Failure of Neoliberalism

When Soviet communism collapsed of its own weight, political and media elites chirped incessantly about the triumph of capitalism. The winning ideology, that of the market, marked the end of history: no more conflicts, at least no more disputes about the rightness of capitalism, would arise. Adam Smith won, Karl Marx lost, end of story.

That the Marx of the Kremlin lost there is no doubt. But the obverse—that capitalism "won" in an absolute way—is erroneous. Capitalism and its core mechanism, the marketplace, are powerful systems of economic exchange. But almost all markets are attenuated by social interventions that protect ordinary people from the bad aspects of markets, and especially imperfect markets. The countries that have created numerous such interventions, regulations, safety nets, social investment programs, and the like are the most successful—Northern Europeans in particular. The United States, the paragon of capitalism and its loudest champion, has quite a bit of social protection as well.

In America, such protections have been the fulcrum of debate about economic policy ever since the New Deal. The entire reason for government intervention was because markets, by themselves, not only leave many people destitute but swing wildly from prosperity to depression. So the New Deal, the stabilization of financial markets, and similar measures gradually grew to make capitalism fairer and less volatile, without stifling one of its major advantages, which is innovation. These actions to bring some order to markets, enforced by public institutions and buffered by social protection, were taken not only in the United States, but, after the Second

World War, in Europe and in much of the rest of the world as well.

That the Europeans, with their very extensive social protection programs—such as publicly paid and high quality health care, solid public education, and so on—have done it better than we have is well known. Quality of life, fairness, and other measures of social well being are considerably advanced in most of Western Europe. Here, since the Reagan onslaught on social protections, household income has stagnated, protection has been dismantled (except to protect corporate interests, of course), and inequality has dramatically increased. Through fewer regulations and social spending, the "magic of the market," as Reagan and his lesser imitators have intoned, will produce much greater wealth for everyone and solve, somehow, social problems as well. That this has not happened could not be more apparent.

What is less well known is how this market mantra has also come to grip the major agencies that are meant to help poor countries. The World Bank, the International Monetary Fund (IMF), and the World Trade Organization (WTO) were created in the aftermath of the Second World War to help war-devastated economies revive and stabilize world capitalism. The Bank would lend money or make grants for reconstruction and new development; the IMF was to help debt-ridden countries and weak currencies; and the WTO was set up to lower trade barriers. Although there were hitches, these global institutions worked as intended and achieved many genuine successes.

Then Reagan and Margaret Thatcher came to power, and decided to inject their free-market fervor into the global economy as well as their own countries'. Because the Bank and the IMF are dominated by the United States (especially so with strong support from the United Kingdom), the agencies took sudden turns to the right. The results for the weak and poor countries of the world were most unfortunate.

The main thrust of the new policy was so-called structural adjustment programs, or SAPs; under their terms, to qualify for Bank loans countries had to enact economic reforms that followed closely the free market model. Some of these countries, mainly in Africa, Asia, and Latin America, and later in the old Soviet Empire, did run inefficient and crony-infested bureaucracies. But the medicine was stronger than it needed to be, and many of the patients in effect expired.

The reasons for this failure are complex, but boil down to the absence—the elimination of—protections both for people and for economic enterprise. Governments were expected to reduce their budgets drastically, which often cut into social protections, health, education, even security. Protections for businesses, co-ops, and traditional ways were devastated. Open trade ideology meant foreign investors—the well heeled of the first world—could buy up the more profitable assets, usually natural resources. Large-scale unemployment, social disruption, and impoverishment followed. Agrarian lifestyles were undermined, and farmers and their families poured into the cities to seek work. Young men were uneducated and unemployable, social authority was disintegrating, and criminality increased.

It is not difficult to understand why the policies failed for the most part. These countries were too weak or underdeveloped to withstand the forces of globalizing capitalism, had few tools at their disposal when things went wrong, and were often beset by ethnic or other latent conflicts that made everything more difficult. As a leading scholar (and former World Bank vice president) Joseph Stiglitz explains, "The result for many has been poverty and for many countries social and political chaos." And in the few places that went through the wringer and came out with some economic growth, Stiglitz notes that "benefits accrue to the well off, and especially the very well off—the top 10 percent—while poverty has remained high, and in some cases the income of those at the bottom has even fallen." Sound familiar?

What is so remarkably daft about these policies is that they ignore everything we've learned about economic growth and prosperity from our own experience in the United States and Europe. In these "advanced" countries, protections for industry and banks as well as protections for families and individuals were gradually but firmly emplaced, and this made it easier to withstand the competitive tumult of the global marketplace. No social protections and few for business were allowed in these developing countries in the 1980s and 1990s. And despite the broad acknowledgment by economists that SAPs failed, they remain in one form or another the reigning ideology of international development.

The result for the developing countries is tragic on many levels. People have been migrating in enormous numbers to find work. International remittances—the money sent home from emigrants—are equal to $150 billion annually, three times the size of official development assistance. Natural resources are mined, pumped, and cut at an alarming rate, with only marginal benefits for the country exploited; the ecological damage is sometimes horrifying. The insistence on export-led growth resulted in agricultural sectors reengineered to produce crops for the lucrative northern markets, but not for self-sufficiency, and when things went wrong with the export market, there was no recourse and famines have sometimes resulted. Public health systems were undermined just at the time the AIDS epidemic grew.

Armed conflict may have been stirred by the social disruptions, lack of security, opportunities for predation, and a new pool of recruitable young men. In short, the SAP experience was a debacle in many places, and fostered inequality and social disintegration in others.

Was this avoidable? Neoliberal economics (the *liberal* comes from the original political meaning of the term, which was applied to those like Adam Smith who argued against feudal or mercantilist restrictions) has its uses, of course, and when markets work well they are usually preferable. But economic development in the global south was a poor laboratory for such policies, particularly because older development models were not failing. The factors that made the United States and Europe strong economically took decades or even centuries to evolve, but these developing countries, with far fewer resources, were expected to do so overnight through unprotected markets.

The neoliberal policies, the work mainly of Reagan and his cohort, also increased debt. "The accelerating magnitude of debt for the most heavily indebted nations is staggering," notes an academic paper on the topic. "In 1970, the fifteen heavily indebted nations had an external public debt of $17.9 billion—which amounted to 9.8 percent of their GNP. By 1987, these same nations owed $402 billion, or 47.5 percent of their GNP. Interest payments owed by these countries went from $2.789 billion in 1970 to $36.2 billion in 1987." A large number of them were in Latin America, and many suffered serious civil unrest or conflicts in this period or soon after. Algeria, Congo, Ivory Coast, Nicaragua, Mexico, and Bolivia were among them. And while some debt relief has occurred as the result of tremendous pressure from the churches and celebrities like Bono, many remain heavily indebted, paying more for odious loans than they can on helping to develop their own economies.

Of course at the same time that the United States insists on unprotected economies in the third world, we continue to protect or subsidize many of our own, and this hypocrisy (Europe does the same thing) is especially striking given the persistence of the market mantra. In the end, of course, it leads one to the inescapable conclusion that marketization has little to do with a vision for global economic growth and everything to do with a strategy for economic enrichment of a few wealthy individuals, companies, and countries. That is a tragedy—real enough for our own stagnant middle and lower classes, and even more so for the billions of people in the global south who were sold, or forced to buy, a set of policies that has been ruinous, unfair, and permanently damaging. The moral deficit we have created might never be remedied.

6   Blood for Oil

It was almost quaint, the demurring in the White House, Pentagon, and the news media at the outset of the Iraq war in 2003. "We won't take forces and go around the world and try to take other people's oil," said Don Rumsfeld. "That's not how democracies operate." The Bushies had learned from their frankness of a dozen years before, when the likes of Dick Cheney, then Rumsfeld's predecessor as secretary of defense, and Secretary of State Jim Baker openly acknowledged the central role of oil in the previous American military invasion of Iraq. President Bush has constantly denied that the invasion was ordered for anything other than the nuclear and biological weapons threat Saddam supposedly posed. Major news media outlets—most notably, the *Washington Post* editorial pages—described those who linked oil and the war as misguided Bush haters.

Some frankness did finally slip out. According to foreign news reports, Deputy Secretary of Defense Paul Wolfowitz, speaking in Singapore in the summer of 2003, spilled the beans. "Asked why a nuclear power such as North Korea was being treated differently from Iraq, where hardly any weapons of mass destruction had been found, [Wolfowitz] said: 'Let's look at it simply. The most important difference between North Korea and Iraq is that economically, we just had no choice in Iraq. The country swims on a sea of oil.'"

For anyone to deny the link between Iraq's massive oil reserves and U.S. interest in the country is a plain act of deception. It simply cannot be described in any other way. The access to oil is one of the sturdiest pillars of American foreign policy. The U.S. support for Saddam Hussein in the 1980s—the infamous Reagan "tilt" during Saddam's war against Iran—preserved such access. The 1991 war against Iraq, after Saddam turned on his patrons, was openly waged to protect Saudi oil fields and to retrieve the Kuwait fields that Saddam had occupied. All of this was an open secret, although what was not fully appreciated at the time was the vast extent of support for Saddam by Reagan, the $5 billion in credits, political credibility, transfer of militarily useful equipment and technology, and valuable military intelligence that saved his regime.

Resource wars are hardly new, of course. The gain of territory and treasure is an old excuse for bloodshed. Everybody does it, or did it, in the big power club. Britain and France certainly turned over a few countries in Africa and Asia, shook them down, and left them to get over it. Spain had its way with most of Latin America. Resources come in all shapes and sizes—minerals, land, timber, knowledge. But few resource warriors do it quite as much as the United States does today, and for black gold it is positively gaga.

The wars in the Persian Gulf have an especially tawdry look to them. We let Iran and Iraq battle it out for eight years while one million died, "balancing" just enough to make sure Saddam did not lose—hence the Reagan tilt to Iraq. We pushed his occupying armies out of Kuwait to protect the al-Sabah and al-Saud royal houses and their oil, but left the Shia and Kurds to languish. We drummed up phony excuses to invade in 2003, all the while denying the relevance of oil. How many people have died in these affairs, and how many Arabs and Persians loathe us because of them, is incalculable.

When U.S. forces took Baghdad in May 2003, looting broke out all over the city. (One suspects that some of that may have been a convenient way to get rid of inconvenient files.) But the oil ministry was protected. The writing of the constitution was done in such a way as to privilege the Iraqi allies of the United States, to give them more of the oil. And the disposition of that oil, ultimately, is a key to understanding the motivations as well. While arrangements are not clear at this writing in early 2006, the control of oil will very likely have American hands on it.

Interventions in Mexico in the early 1900s and much of Central America throughout the century; Angola and Chile through covert action in the 1970s; threats vis-à-vis Iran and Iraq in the 1950s; the Philippines in 1905; and Vietnam and Southeast Asia from the 1950s to 1970s, all involved resource wealth. Venezuela, already a target of covert action in this decade, is a likely future victim.

Today, as Michael Klare has described in detail, U.S. military operations are ready to intervene to protect American access to Caspian Sea oil, which is in a dodgy neighborhood in the Caucasus, north of Iran.

The "blood for oil" habits are a chronic American syndrome of wasteful living at home, belief in our right to own other countries' resources, and willingness to act militarily to extract those resources when need dictates. The fact that the military interventions, support for noxious allies, covert action, and the like are usually couched in more acceptable language—anticommunism, anti-terrorism, etc.—does not mask the basic intentions. One of the most starkly hypocritical debates in America over the last twenty years or so has been about energy and petroleum dependency. Scratch the surface and there are two fundamental realities: our wantonly profligate use of energy, and our moral assumptions about getting it from others. Those are the twin pillars of the "blood for oil" ideology.

There is no illusion abroad about this ideology. A British columnist put it succinctly in late 2005: the "irresponsibility" of U.S. policy was apparent "at the very time when U.S. dependence on oil imports has been spiraling, consumer profligacy has also been sharply increasing," and that "the danger is that Bush's response will be to become even more interventionist in those countries which underpin the supply." Or consider this from a prominent Arab analyst on the eve of the invasion of Iraq in 2003:

> On all the major issues concerning the oil market, from raising prices to equitable levels and improving the conditions for production quotas to the nationalization of oil supplies, Iraq had always been among the hawks in OPEC. As a matter of historical record, therefore, Iraq has always presented an obstacle to the U.S.'s oil-market strategy. This explains why the U.S. administration's behavior towards that country is so implacably vindictive, and why, in the process of occupying Iraq to drive oil prices down to the cheapest possible levels, it wants to drive a lesson home to all nations opposed to the U.S. and use the fate of Iraq as an example to intimidate all developing nations.

Whether every word of such analysis is true is not at issue. More important is how it reflects opinion globally about the blood-for-oil nexus. There are no illusions, no self-deception, about American intentions in Iraq or Venezuela or the Caspian. It's about oil. It always has been.

7   Agribusiness

Agriculture in this country, long a point of pride for its remarkable productivity and agrarian ethos, now resembles a cartel that is damaging the world's small farmers and not doing much for the health and well being of all other Americans, either.

The U.S. government subsidizes American farming to a significant degree, and most of that money—$20 billion or so annually—goes to four large agribusiness corporations. This has been a bipartisan boondoggle for many years. There have been attempts to reduce the harm of farm subsidies, but the system rolls on, only the details change. If anything, the program increasingly favors the wealthy agribusiness interests over small farmers, all but excluding new entrants. The ecological damage to soil and water is staggering. And the effects on farmers in the developing world—where we often dump underpriced, subsidized foods—are devastating.

One of the most fascinating aspects of farm subsidies is how they affect health. Consider corn. Over the last ten years, Washington has fed corn farmers nearly $50 billion in subsidies, and there are places in the heartland that are piled high with millions of bushels of corn that may never be consumed. What cheap prices for subsidized corn have done, among other things (like hurting Mexican agriculture and sending some of their farmers north), is to create an incentive for all processed food manufacturers to use corn syrup as a sweetener instead of other sugars. It is high fructose corn syrup, and if you look at the ingredients of just about any soft drink and many other foods, it's in there. And it's unhealthy. Since 1970, shortly after it was developed, the use of this sweetener has boomed by 1000 percent. It is now strongly correlated with obesity, and may have particularly damaging effects on children. It could be linked to what medical specialists describe as an epidemic of diabetes in the United States, an extremely costly consequence that dwarfs whatever savings it gains for consumers.

The high-fructose corn syrup problem—even Paul Newman's lemonade has it—is just one aspect of health-related problems in farm subsidies. Genetically modified foods are sure to be one of the over-the-horizon health problems we must deal with, and the United States is a leader in allowing GM foods. Already GM seeds are required to be used in some foreign assistance programs that require local farmers to buy the seeds year after year (because crops do not produce them naturally) at high prices. This is also damaging to crop diversification. It helps the agribusiness owners of the patents, however.

Food policy driven by the giant businesses also creates a major problem via the reduced prices for commodities that are then dumped on developing countries. In fact, many of them are grown—and subsidized—specifically for export. Much of the U.S. budget for food assistance to poor countries buys up excess production, yet, as economists know, this is a highly inefficient way to deliver aid. It costs more, uses more energy, and ruins local prices for small farmers in Africa, Latin America, and other places where development and food security have an intimate, and unwarranted, relationship.

According to a World Bank study, agricultural subsidies cost the developing world $350 billion annually. Official development aid—foreign assistance—amounts to only one-seventh of that total, or $50 billion. In this, Europe is every bit as guilty as the United States. Keep in mind that some 800 million people worldwide live on the brink of starvation. One would think that subsidized grains might help that situation, but the system of production, sales, transport, and its effects on local agriculture are so distorting that it does not really address food security.

The subsidies extend to nonfood items like cotton, which earns $280 billion in subsidies and punishes the 10 million people in poorer countries that depend on the crop.

These well-known travesties violate the obsessive insistence on free market practice and the proud declarations of how America feeds the world. U.S. foreign assistance—actual loans and grants for development—come nowhere near the cost of agricultural subsidies, measured either in taxpayer outlays or their effects on small farmers in the global south. And it can be said without a moment of hesitation that as long as these kinds of subsidies and protectionism persist, all the pontificating about free markets from our political and opinion elites is just hypocrisy, plain and simple.

The price we pay in health impacts, environmental degradation, and losses in farm communities can be added to the wicked effects on the poor of the world. As a major study of the Institute for Agriculture and Trade Policy puts it, "Hunger is not inevitable. Malnutrition is not a consequence of food scarcity, but a result of the way economies are organized and of political choices to address—or ignore—the causes of hunger. In the twenty-first century, we have the means to defeat hunger: we grow enough food, we know enough about redistributive economics, we have the political tools to ensure inclusive decision-making and we can afford to provide the basic needs that protect every person's entitlement to an adequate, nutritious diet."

By selling food to the hungry, food that is so grossly subsidized that it costs three times its true value, we undermine food security, global stability, and our own sense of principled action in what should be our most generous gift to humanity, and to ourselves.

8   The Reagan Doctrine

Frustrated by what they saw as Soviet advances in Africa, Central America, and Central Asia, the right-wing cohort around Ronald Reagan renewed an idea for countering communism, what came to be known as the Reagan Doctrine. The logic was simple. Just as the Soviet leader Leonid Brezhnev and Chinese leader Mao supported armed insurgencies against colonial or U.S.-aligned states, American power would now encourage and support rebels against communist states. The results have been a catastrophic failure.

There were three targets in particular: Afghanistan, Nicaragua, and Angola. In Afghanistan, a Soviet-backed regime had come to power in 1978 after a coup, and to support this tottering government, Soviet forces invaded and occupied the country in December 1979. In Nicaragua, a widely popular revolution by the Sandinistas, named for a legendary rebel, Augusto César Sandino, overthrew a decades-old oligarchy in 1979, and moved gradually toward the communist orbit. And Angola, an oil-rich Portuguese colony, won its independence in 1974 after the right-wing regime in Lisbon collapsed; the Popular Movement for the Liberation of Angola (MPLA), which had waged a war of independence for twenty years with support from Cuba, came to power. (Cambodia is often cited as a fourth example, but that was more a legacy of the Vietnam War.)

In Angola, a rival rebel group led by Jonas Savimbi vied with the MPLA for post-independence leadership, but when several peace agreements failed to hold, Savimbi returned time and again to the battlefield to attempt to settle matters by force. Savimbi was supported by the white apartheid regime in South Africa—telling enough—and then by Reagan. By turns a Maoist and a "freedom fighter," Savimbi was a classic warlord who violated several peace agreements. The Organization of African Unity, the continent's main multilateral body, condemned him as a tool of South Africa. Reagan welcomed him to the White House and hailed him as "America's best friend in Africa." He provided the warlord with $30 million annually in weapons and other needs to sustain the insurgency.

When the war finally wound down—Savimbi was killed in an ambush in 2002—its toll was staggering: an estimated one million were dead, many of them civilian collateral deaths of the war, including 300,000 children. The MPLA stayed in power, Angola was crippled, and the surrounding countries also felt the impacts of twenty-seven years of warfare. "The effects of civil war led to Angola having the highest infant mortality rate in the world in 1990—of every 1000 children born, 350 died before the age of five," reports an Australian health journal. "Children who do survive are affected by war in other ways including poverty, malnutrition, separation from parent(s), exposure to destructive violence, witnessing death or other atrocities, permanent disablement, having parents who are seriously affected by war experiences, and the loss of life-sustaining infrastructure of society." When the war finally ended and relief workers were able to access areas controlled by Savimbi, they described the malnourishment and disease in those areas as "catastrophic."

In Nicaragua, the United States was faced with the repudiation of its thirty-year history of support for the Somoza family of dictators when the popular Sandinista movement, led by Daniel Ortega, overthrew the last of the Somozas in 1979. As with Cuba, U.S. hostility hastened the new leftist government toward the Soviets. When Reagan took office, the Sandinistas were high on the right-wing hit list. With covert and largely illegal support for an oddball collection of Somoza partisans, mercenaries, drug runners, and anticommunist zealots—collectively called the *contras*—an immense amount of pressure was exerted on the Managua government. The *contras* "were major and systematic violators of the most basic standards of the laws of armed conflict," concludes a Human Rights Watch report, and supported themselves by building cocaine networks that persist to this day.

The Central American presidents consistently voiced their opposition to Reagan's lavish support for the *contras*, to no avail. And so did Congress, but the Reagan White House circumvented congressional prohibitions to funnel arms and other aid to the *contra* gangs. Carried out by the notorious Lieutenant Colonel Oliver North, convicted for his misdeeds (but now a host on Fox News), the illegal aid to the *contras* prolonged the war that killed 50,000 and displaced 200,000, by conservative estimates.

By 1990, exhausted by the isolation and *contra* violence, and having mismanaged the economy, the Sandinistas lost an election to a moderate coalition. The Soviets never had much interest in Central America and jettisoned support for Ortega early on. The Sandinistas have remained close to power, however, in a country—and a region—which was devastated by Reagan's war.

"The Contra war left Nicaragua bitterly divided and heavily armed. An estimated 25,000 to 100,000 weapons remain in civilian hands," reports the U.S. Library of Congress country series. In the early 1990s, the *recontras*, as they called themselves, carried out "kidnappings of Sandinistas for ransom and attacks on members of farm cooperatives. In 1993 the United States Department of State described their activities as principally criminal, with political overtones." A Senate committee in 1989 reported that "there was substantial evidence of drug smuggling through the war zones on the part of individual Contras, Contra suppliers, Contra pilots mercenaries who worked with the Contras, and Contra supporters throughout the region." The drug smuggling problem in the United States in the 1980s and subsequently is largely facilitated, if not run by, the networks supporting the *contras*. From the beginning, the criminality of the *contras* was breathtaking.

Afghanistan represents the most vivid assertion of success by Reagan Doctrine advocates, who say that the CIA's supplying of the mujahideen defeated not only the Soviet army but precipitated the collapse of Soviet communism itself. These claims are so insistent perhaps because the actual consequences are so obviously disastrous.

It was Carter national security adviser Zbigniew Brzezinski who first suggested that the United States should provoke the USSR into invading Afghanistan (with a threatening movement of U.S. forces into the region). But the Soviets were rattled mainly by the Islamic revolution in Iran and parallel unrest in Afghanistan, then friendly to Moscow but not wholly dominated by the Soviets. After they did invade and occupy Afghanistan in late 1979, a gradually effective guerrilla resistance formed, generally called the mujahideen, a mixed bunch but significantly "funda-mentalist" Muslims waging jihad against the Russians.

The CIA funded the mujaheddin, supplying training and weapons in the early 1980s and more openly providing Stinger antiaircraft weapons and the like after 1985. The Soviets, sustain-ing some 14,000 casualties, did withdraw in 1989. But the U.S. actions were only part of a much larger set of events and consid-erations for Moscow. Very early in the occupation, the Soviets had misgivings about the Afghan venture and were seeking a with-drawal strategy, as Politburo minutes show. Gorbachev, who came to power in 1985, was committed to withdrawal from the start of his premiership. So the major part of U.S. arms began to flow only *after* the Soviets had begun their maneuvers to get out. Since the Soviet public knew little of what was going on in Afghanistan, the notion that this misadventure caused the downfall of the Soviet state is not credible.

What is plain is the devastation the war wrought, however, and not just in the endlessly beleaguered Afghanistan. Weapons shipped to the mujahideen were discovered during the 1980s and 1990s in civil wars as far away as Mozambique. An estimated 3 million AK-47s were provided to the muj. That fighting force itself became the beginning of the jihadists, including Osama bin Laden's debut as an anti-West fighter, who have since grown to be such a significant threat worldwide.

Afghanistan was thoroughly undone by the long wars and then the nearly total abandonment by the United States after the Soviet forces departed. It became the quintessential failed state, its economy buoyed by heroin production, its society radical-ized or newly repressed by warlords and religious zealots. After a lengthy civil war, the Taliban came to power in the late 1990s and were soon seen as among the most backward regimes on the planet. President Bush nonetheless sent them aid in the summer of 2001.

So it is obvious that the Reagan Doctrine was not only a colossal failure, but immoral by nearly any standard of international conduct. It is almost impossible to calculate its bloody consequences—millions dead, many more millions displaced and driven from their homes, millions of children orphaned, maimed, and left destitute; failed states and devastated economies in place; cocaine and heroin industries flourishing; international criminal gangs and terrorist networks (*contras,* al Qaeda) given sanctuary and official standing, some now still thriving. A truly astonishing range and depth of destruction.

It would be difficult to conceive of a more depraved set of policies. The intentions were illegal and immoral from the outset, and the results have cascaded into one series of catastrophes after another.

9   The War in Vietnam

From the early 1960s to April 30, 1975, the United States government pursued its most misbegotten war of the last half century in the small, faraway land of Vietnam. More than one million Vietnamese were killed in the conflict, mostly due to U.S. bombing, and more than 58,000 Americans, almost all soldiers, also perished. The United States was defeated by the Vietnamese, and Vietnam to this day remains firmly in the hands of the Communist Party.

The war was foolhardy from the start and many critics said so at the time. It shook America—its politics and culture—as few wars have. It also left an indelible mark on Southeast Asia, America, and the world.

As is often the case, the war was rooted in other wars. The Second World War left Vietnam in disarray; a French colony occupied by the Japanese, Vietnam gave rise to the national liberation movement led by Ho Chi Minh. The French tried to reassert colonial control but met with stronger and stronger resistance led by Ho, culminating in a decisive defeat at Dien Bien Phu. Vietnam was divided between north and south, with the north controlled by Ho and the south beset by a series of unstable regimes backed by the French and the United States. A guerrilla war in the south reignited the country, and the United States became increasingly involved under President Kennedy. By 1964, it was a full-scale war and President Lyndon Johnson had made an enormous commitment to prevent a communist victory. A suspicious incident in the Gulf of Tonkin led to congressional approval of a much larger U.S. intervention, and before long, 500,000 American troops were fighting in the country.

Historians argue whether the United States could have negotiated an early end to the war, what the terms might have been, and how the conduct of the war affected this diplomatic effort. In the end, what did result was an astonishing scale of devastation, no meaningful peace agreement, and a communist victory. The peace movement in the United States (and Europe) grew to massive size, affecting culture and social relations as well as politics, and hastened the U.S. retreat.

The effects of the war outside Vietnam have been felt for decades. Not only did the war destabilize Southeast Asia, with Cambodia victimized more than any other neighbor, but the U.S. defeat was used as an excuse to bolster military regimes and dictatorships in the Philippines, Indonesia, and elsewhere. One "exit strategy" was to arm the South Vietnamese in lieu of sending more U.S. troops, the so-called Nixon Doctrine, which actually had its largest effect in Iran in the 1970s, where U.S. military support for the shah led to a cascade of misadventures we suffer from to this day.

Vietnam also created a myth about American "humiliation" (and the supposed gains of the Soviet Union), and gave rise to a revanchist political pathology in the United States that resulted in the election of Ronald Reagan. Because it was America's first clear military defeat, it created a lasting sore in the body politic, which continuously plays out in politics and the right-wing news media.

One could, instead, view the defeat as a lesson about arrogance, misunderstanding foreign political cultures, choosing military solutions over nimble political maneuvers, and wasting American assets (and many others' lives) on a decrepit and rigid anticommunist ideology. The "humiliation" was a lifesaving lesson, or could have been. Iraq is proof that it was not.

Above all, the Vietnam War demonstrated how deadly and counterproductive military containment of the Soviets and Chinese often was. In Southeast Asia alone, the cost has been in millions of lives lost, and many millions more wasted in the ruins of Cambodia, Laos, Indonesia, and elsewhere. The belief that a line needed to be drawn in blood was a costly mistake repeated throughout the world over a period of decades, and it remains as a legacy both poisonous to our politics and lethal to much of the world.

10   The Waltons Go Global

The Walton family of Arkansas, owners of Wal-Mart, are the wealthiest people in the world, owners of the largest retailer and America's leading employer and moneymaker. It has been a remarkable success story, and it is remarkable not only in the astonishing wealth of the family—$84 billon—but in how it has altered business practices in the United States and has quickly become the paragon of the globalized economy.

At the same time, Wal-Mart's predations are stirring one of the most interesting and useful debates we have had about poverty and a "living wage" in America. Since average incomes have stagnated over the last thirty years, it's an immensely important discussion, with profound implications for the rest of the world.

The critics of Wal-Mart make a simple case: its practices demand the lowest prices for everything, from overseas labor to in-store labor, and this not only leaves those workers wanting, it also undercuts local businesses that cannot compete with this global behemoth. The Wal-Martification of America is well under way, with the superstore dominating everywhere it goes, particularly in the smaller markets where it has little competition and less of the onus than it does in major urban areas.

The firm's defenders make two points. First, low prices are good. Poor people shop there precisely because of low prices, which make it possible for them to get by. So there's a vast social benefit, no matter how unintentional. This is the way of the world and the local mom-and-pop stores would lose out in the long run anyway. Second, people line up for jobs at Wal-Mart despite the supposedly low pay, long hours, and scant benefits.

It's not difficult to rebut the defenders, who read from Wal-Mart's war room script, but the points they make are at the heart of the globalization debate generally, so they must be taken seriously. What's at issue in the Wal-Mart fight pretty much encompasses what's at issue with regard to economic globalization as a whole.

The mantra about low prices overlooks the fact that Wal-Mart is rich enough to afford better wages and benefits at home and abroad, while keeping prices low and their own profits ample. In this, there is a sense of fairness that runs through the entire discussion of overseas outsourcing, imports, and low-paying service and clerical jobs here.

The criticism that emerges is that this model will drive everyone—other retailers, large and small—to the bottom. "The company's business practices, such as leaving more than half of its employees not covered by its health insurance plan, also contrib-

ute to real reductions in the purchasing power of its empl
Christian Weller, senior economist at the Center for Ame
Progress, wrote in response to a pro–Wal-Mart column in the
Washington Post. "And when many companies follow Wal-Mart's
lead, as they must, millions of Americans are left with declining
real wages and rising debt—exactly what is happening in today's
economy. If Henry Ford wanted his workers to be rich enough to
buy his cars, Wal-Mart is leading us to an economy in which its
employees are barely able to shop at Wal-Mart."

A similar line of reasoning applies on the jobs issue. The *New
York Times* columnist John Tierney, citing the same Wal-Mart
sponsored "research" as did the *Post*, asserted that "Wal-Mart has
been one of the most successful antipoverty programs in America.
It provides entry-level jobs that unskilled workers badly want—
there are often five or ten applicants for each position at a new
store." Other retailers on average pay a lot more; Costco, another
discount store, pays more than 60 percent higher, and its workers—
surprise!—are more productive. But the fundamental fact is that
the large number of people in line for Wal-Mart jobs is a comment not on Wal-Mart, but on an economy that cannot produce
decently paying, secure jobs with sufficient benefits, like health
care for children.

"As of last spring," one report noted in late 2004, "the average pay of a sales clerk at Wal-Mart was $8.50 an hour, or about
$14,000 a year, $1,000 below the government's definition of the
poverty level for a family of three." And those long lines for jobs
may be occasioned by the exceptionally high rate of turnover, more
than 50 percent annually (compared with half that at Costco).
Discriminatory practices against women have been particularly
severe, and the company is known for its vicious union busting.

Because pay is so low, the U.S. taxpayer ends up footing quite a
bit of the shortfall. A congressional committee report in 2004 calculated the following: for a Wal-Mart store with 200 employees,

"the government is spending $108,000 a year for children's health care; $125,000 a year in tax credits and deductions for low-income families; and $42,000 a year in housing assistance." That is more than $2,000 per employee, or a *yearly cost of $2.5 billion* for Wal-Mart's American employees, and that is only the federal outlay. In 2004, Wal-Mart made profits of more than $10 billion.

It is not just what happens at Wal-Mart that matters, however, but also the company's impact on the business world. When an industry like retailing—an enormous segment of the entire U.S. economy—becomes so concentrated, new entrants are at such a disadvantage that no one can get in, apart from niche players. And the largest impact of all is on its suppliers. Some say that Wal-Mart's tightfisted policies have made the whole of American business more efficient, as with just-in-time inventory controls. But by all accounts, Wal-Mart is so predatory in its purchasing power—unrivaled purchasing power—that it dictates not only wages, but prices and product development. If a supplier of lingerie or bicycles or food or jeans does not meet its severe demands, Wal-Mart will go elsewhere, usually to China, and leave the American companies in the lurch to import, to lower wages or quality, or to die.

And those overseas suppliers? As one might imagine, they are under the same relentless pressure that American suppliers are, and they have fewer labor or environmental laws to constrain their drive to the bottom. One Wal-Mart manager described what he found when he was tasked with the Central America operations: "factories whose fire doors were padlocked from the outside, and where women workers were fired if they turned up pregnant." He also discovered that Wal-Mart suppliers in Honduras were locking workers behind barbed wire for ten to thirteen hours a day. When he reported these things to the Arkansas headquarters, he was fired.

The scale of Wal-Mart operations abroad is vast. It imports

10 percent of all Chinese-made goods into the United States. And 80 percent of the items it sells are now made in China. Wal-Mart has more than 6,000 suppliers worldwide, and 400,000 employees in its 2,400 foreign stores. Its practices overseas are among the most notorious of any major American corporation. A class-action suit brought in 2005 by garment workers at factories in Bangladesh, China, Indonesia, Nicaragua, and Swaziland, detailed how Wal-Mart "failed to ensure compliance with its code of conduct for suppliers and misled the American public about its efforts on behalf of foreign workers." The stories abroad mirror those at home: labor exploitation, environmental assaults, union busting. The model has been perfected, why not apply it everywhere?

And this is the nub of the problem for the rest of the world. Just as Wal-Mart has had a powerful downward effect on personal income and quality in the United States, it will likely have those effects in its foreign operations. At the same time, the "model" is all about the lowest possible wages. People may line up for jobs in China or Thailand, too, but sweatshops have always found people who had to have a job, any job, to survive. As America's largest company, Wal-Mart has this deeply damaging effect all over the world, the business model in which the poor are linked to the poor in a giant network of unjust dependencies. The Waltons, however, have made out just fine.

11   Gilded Democracy

Scholars sometimes refer to certain kinds of political systems as "guided democracies," those that have some features of democratic governance but are largely in the hands of some powerful force. Turkey is a guided democracy, controlled by its overweening military. Iraq is guided by the U.S. occupation. Examples abound.

Our own democratic institutions in the United States are strong and mostly independent from one another, but we suffer from a particular distortion of democracy, a "guide" that affects our own lives and those of many around the world. That is the perverse influence of money on politics. It shapes trade, environmental action, war—virtually every aspect of our global presence. You might say we have not merely a guided, but a *gilded* democracy.

Lobbying by China, to cite one fairly obvious example, can sometimes take your breath away—or your senator's judgment. China and Hong Kong have spent $20 million over the last ten years on lobbying Congress to relax textile imports and similar measures. As the trade deficit with China grows, and Beijing owns more and more of our treasury bond debt, the matter of how the Chinese use high-octane K-Street firms to do its bidding is of more than incidental interest. (Despite Tom DeLay's diktat ordering those seeking favor to use only Republican lobbyists, the lucre remains somewhat bipartisan.) It has been estimated that the vote of a member of Congress can be had for about $3,000 in campaign contributions and other such goodies, so $20 million can go a long way, and it has.

Consider the words of Representative Rahm Emanuel of Illinois, who as a veteran of the Clinton administration should know a thing or two about the influence of money: "When the Speaker's gavel comes down, it's intended to open the People's House, and lately it's looking like the Auction House," he told CNN's Lou Dobbs. "Whether it's an energy bill that gives more than $8 billion to the oil and gas interests while oil's at $64 a barrel, whether it's a corporate tax bill solving a $5 billion problem with a $150 billion solution, whether it's a pharmaceutical, prescription drug bill where the industry gave $132 million and walked away with $135 billion in additional profits."

And while Congress is the traditional target of choice for much lobbying activity, the White House is famously vulnerable, too. The Center for Public Integrity, a leading research group on the influence of money in policy, lists the top conflicts of interest of the current occupant of 1600 Pennsylvania Avenue: "Andrew Card, the president's chief of staff, previously lobbied on issues like product liability for the American Automobile Manufacturers Association. Similarly, Philip Cooney, former chief of staff of the President's Council on Environmental Quality, lobbied for the American Petroleum Institute (he resigned his post last June [2005] after revelations that he had altered national climate-change reports and will soon be heading for a position with ExxonMobil). And there is Edwina Rogers, who was a lobbyist for NASSCOM (India's IT trade association) before working as associate director for the White House's National Economic Council. Then, in 2004, Rogers left government to lobby for the Erisa Industry Committee, which represents the interests of employers in matters relating to retirement, pensions, health care and other worker benefits."

In one short passage, we see how the tendrils of industry interests and governance intertwine. The Center identifies fifty-two major lobbyists who also served as Mavericks, Pioneers, or Rangers in the Bush campaigns—the high-dollar fund raisers, like Jack Abramoff, a Pioneer (more than $100,000) who represented some nineteen clients with the White House.

There is little that is secretive about any of this. Arnold & Porter, a major lobbyist in D.C., advertises on its Web site: "Defense and National Security, including cybersecurity, export control, Foreign Corrupt Practices Act, homeland security, bioterrorism and electronic surveillance issues, among others, led by a team of former General Counsels of the Department of Defense, the Central Intelligence Agency and the National Security Agency and former senior officials from both the Justice and State Departments."

Corporate lobbying touches every aspect of U.S. policy making, and some—like environmental rule making and laws—have profound, lasting effects on the rest of the world. The business sector has worked tirelessly to defeat positive action on climate change, pesticide exports, toxic waste dumping, and hundreds of other practices that despoil the earth and endanger human health around the globe.

According to Political MoneyLine, corporate lobbying reached $2.14 billion in 2004, including the lobbying of Congress and 220 other federal agencies. These figures may shock and awe, but they are really commonplace. We see the headlines about Abramoff and DeLay and countless others every day. But how does it affect our role in the world?

Some well-known cases illustrate the kinds of influence lobbyists have—China trade, support for Israel—but the scope of lobbying's tawdry effects goes much further. The Abramoff and DeLay scandals have revealed some of that—the former majority leader's posh trips to Russia, for example, were underwritten by a Bahamian-registered company that serves Russian interests, aimed at gaining U.S. aid and favors. This is a run-of-the-mill occurrence that would not have earned notice if it did not involve two of the most corrupt people in Washington, whose overreaching is legendary. But the simple junket, the trip abroad for "fact-finding" is more often linked to a piece of legislation that provides something tangible for a business interest or government.

A number of countries—Indonesia, Turkey, China, and so on—engage lobbyists to improve their image, to overcome bad publicity for human rights violations, massacres, and other things that can produce bad press. This can yield results. Congress is less likely to investigate bad behavior abroad if they are sufficiently buttered up. The news media is not impervious to this, either, and many lobbying gimmicks are aimed at the gatekeepers of the fourth estate.

But it goes the other way too. The U.S. government uses several mechanisms to influence foreign governments—foreign aid, after all, is significantly about economic development. American corporations are relentlessly lobbying foreign governments, and chafe at the antibribery rules the United States more or less enforces. In some smaller countries, American corporations can have profound influence through lobbying, starting in Washington, moving through the public and private international banks, and ending in the capital of the target country that might have a natural resource or utility to sell off.

But lobbying in Washington all by itself reverberates around the world. The aerospace companies like Lockheed Martin shape the entire aviation industry of the world with their clout on defense contracts. Agricultural policy, including who gets food and who does not in the poorer parts of the world, is influenced mightily by heavy-handed lobbyists like Archer Daniels Midland. Health care everywhere is shaped by the hard-bought privileges of the pharmaceutical companies. Because of America's economic size and the reach of its corporations, every soft footfall of favor seeking and bribery inside the beltway is like an earthquake somewhere in the global south.

This kind of corruption is not an American invention. But like so many things, we've perfected it, knowing but intentionally unmindful of its global reach.

12   Nourishing the Seeds of Islamic Militancy

I suppose most readers will say it's a stretch to claim that America created Islamic militancy. But sometimes big claims can yield useful insights. Let me take you through this one.

One of the strategies of the Cold War was to utilize as many right-wing militaristic thugs as we could stomach, which turned out to be quite a few, to fend off the USSR. In the oil-producing regions, this was especially important. From the time of the First World War, when Britain changed the way it powered its formidable navy from coal to oil, the link between petroleum and security has been a fundament of big-power machinations.

As the United States grew into a global power, we took up this imperative to control oil with gusto. It included, most particularly, the protection of the Gulf monarchies, repressive regimes of a traditional kind. This status quo—the Soviets with their own oil reserves and a few friends in the Middle East, and the West with the rest—was acceptable to the global powers.

The price of this became increasing repression in Iran in particular. The CIA coup against the democratically elected government of Mohammed Mossadegh in 1953 was perhaps the key moment in this tale. Mossadegh had threatened to nationalize the oil industry, a common aspiration in the Middle East at that time, and indeed accomplished peaceably by others (including Saudi Arabia). Clearly, national ownership of oil resources did not mean the West would not have access; Iran would need the marketing and technical capabilities of the same oil companies it intended to buy out. But the action was too much for the Cold Warriors, and Mossadegh was thrown out. And, with him, went a viable liberal and leftist opposition to the ruthless American vassal, the shah (whose father, a military man, had seized power in 1920s and made himself "shah").

The absence of political pluralism meant that there was only the shah and his increasingly corrupt cohort of military officers, oil tycoons, foreign courtiers, and SAVAK, his bloodthirsty intelligence service. Opponents of his rule within Iran had to gravitate to the souk, where the imams were primary, who remained relatively untouched by the shah, in part because he did not consider them to be dangerous. (Some, like Khomeini, did flee.) The

United States supported the shah lavishly throughout his reign, overlooking his human rights violations and arming him to the teeth. Nixon and Kissinger, in particular, were near sycophants. U.S. military assistance braced the shah's despotism, and gradually (and invisibly to Western observers) fueled the religiously based opposition at the same time.

By 1978, when the shah was weakened by his excesses, political Islam came of age. The interim government ushered into power when the shah left the country could not withstand the demands from the "street," then filled with militants imbued with Koranic fervor and a new national hero, Ayatollah Khomeini, who returned from exile to lead Iran to the Islamic revolution.

This victory for a politically astute, if reactionary, ideology was not lost on other Muslims opposed to their regimes and to the United States and other foreign domination of their countries. The humiliation of the United States was intoxicating, and militants attempted takeovers elsewhere—Egypt (murdering Sadat), Algeria, Afghanistan, Sudan, Lebanon, and others, more or less succeeding and building confidence and momentum along the way. A similar dynamic has afflicted Israel, where the religious Hamas successfully challenged the secular political establishment of the Palestinians as well as the Jewish state. The U.S. support of the mujahideen to oust the Soviets from Afghanistan is an oft-told tale of short-term gain for long-term pain.

Where the old Soviet-backed strongmen were in charge, such as Syria or Iraq, Islamic radicalism was largely repressed. For how long, of course, is another matter. In Iraq, for example, the U.S. backing of Saddam in his war with Khomeini's Iran was viewed suspiciously by many Muslims. A number of Muslim societies, when permitted a vote, are prone to support the Islamicists, and oftentimes this is in reaction to U.S. policies in Israel, Iraq, Saudi princes, and similar missteps.

So, no, American political leaders did not set out to create a worldwide movement of hatred against the United States. And there are roots elsewhere for political Islam, such as the Muslim Brotherhood in Egypt. But because expediency usually trumped principle in U.S. foreign policy, and political pluralism was suppressed in favor of American rent-a-cops, the result has, predictably, been a growing antipathy for America and its hypocrisies. And the avatar of that antipathy has been militant Islam. We remain deaf to this lesson, at our peril.

13   Spreading the Word

In the mid-1990s at a conference in Boulder, Colorado, the Shanghai bureau chief for a major American newspaper told me in an offhand way that the Chinese regard America as a "Christian crusading nation." I was a little taken aback by the comment, not because I thought it doubtful, but because it rang so true and the intelligentsia in this country had all but ignored its broader meaning. No longer, of course, can one miss this phenomenon.

For centuries, Christian missionaries have tripped over one another to convert the heathens and animists and other misbegotten souls, so there is nothing especially new or American about evangelicalism. Except for its breadth and fervor. The Bible-thumping preachers in the tents on the outskirts of town—the way I first encountered evangelicals in Indiana in the 1960s—are now running the show in Washington and spreading the Word worldwide. The fact that many are allied with the military in some important ways makes this much more than a "let's build a hospital in the Congo" kind of movement. It's about power and dominance, moral certainty and intolerance.

That is not to say that all church meddling in the developing world, the former Soviet Union, and other places—including

Europe—is wicked. The "parachurch agencies" that carry both the Word and relief and development for the poor are doing more good things than bad, but many walk a fine line and more than a few are proselytizers first and development agencies last. As one sympathetic writer describes it, "The first wave of these agencies arriving in the non-western world—groups such as World Vision, Youth for Christ, Full Gospel Businessmen, Women Aglow, Campus Crusade for Christ, Scripture Union, and the International Fellowship of Evangelical Students—looked like a new missionary invasion, and perhaps, some feared, a new form of religious colonization."

The impetus for Christianizing the world in a distinctly American form (including the belief that God has a special mission for America) has gone through ups and downs. But a steady expansion, accelerating in the 1980s, is apparent. Consider the Campus Crusade for Christ. "Our goal for this decade is to help give every man, woman, and child in the entire world an opportunity to find new life in Jesus Christ," its international division says. It has 16,000 staff in sixty countries. It has proliferating "ministries" that include Athletes in Action, Global Media Outreach (whose goal is one billion presentations of the Gospel), Family Life, and dozens more geared to health care professionals, graduate students, legal services, and how-to support networks for evangelists.

Like their homegrown bases, the international movement increasingly stresses personal growth, family life, sports and entertainment, financial planning, and other daily concerns in order to make the ministries attractive and relevant. Just as they fill gaps in our depleted social capital in the United States, they seek to do so abroad.

Campus Crusade and its many offshoots may be among the larger worldwide efforts, but the old-time religion types are going global too. The Pentecostals, one of the most fundamentalist churches, reports world membership at 1.6 million at the end of

2003, nearly 90 percent of them outside the United States. I can recall an anthropologist in southern Mexico describing to me in 1980 the conversion of Mayan tribes in the nearby jungle by Pentecostals, and I thought he was joking. Now I realize that was just a small tip of the iceberg. "God is calling the Pentecostal Holiness Church to become a unique instrument of world evangelization which will penetrate many unreached peoples of this earth with the gospel," says the International Pentecostal Holiness Church's Jerusalem Proclamation. "We are to gear up like an invading army to go into the strongholds of the enemy with our lifestyle of worship and witness in order to spread the kingdom of God and the reign of King Jesus. We are to go on a wartime basis in order to carry out this God-given mandate."

"Excite church members of all ages about the challenge of reaching one of the largest unreached nations on earth," says the International Mission Board site of the Southern Baptist Convention. "India has the world's largest Hindu population. India has the world's second largest total population. India has the world's third largest Muslim population. The people of India need to hear about Jesus." And: "South Asia has the greatest concentration of lostness in the world. Pakistan, India and Bangladesh are among 10 nations with the world's largest Muslim populations. India is the world's largest Hindu nation." You see, the people of South Asia are lost because they are Hindus or Muslims.

The missions now include traditional goals of planting churches, Bible study and prayer groups, as well as colleges, humanitarian activities, television and publishing, business networking, travel, adoption agencies, and so on. Importantly, they are developing local people to be leaders in the churches and in the worldwide movement.

One's view of all this godliness would likely depend on one's view of God. Is God only a Christian who favors Americans above all else and commands his minions to convert the world? If you believe that, then the evangelicals at work in all corners of the globe are probably to your liking. If you believe that God and faith can take many shapes, or if you are not a believer, then perhaps the missionary zeal of these dozens of churches and parachurches is troubling.

The evangelicals do not stop with religion or family life, however. They are increasingly involved in politics at home and abroad. Recent controversies have flared about the growing influence of the fundamentalists in the military, for example. Evangelical groups have driven much of the family planning and HIV/AIDS agenda of the Bush administration, emphasizing sexual abstinence, with potentially catastrophic consequences in areas at risk. They have created a sizable lobby for Christian "solidarity," a kind of hybrid human rights movement for religious freedom (active where Christians are persecuted).

In the Middle East, which many evangelicals consider both the Holy Land and a hornet's nest of evil Muslims, evangelicals are trying to get a foothold by opening churches and recruiting members. Some followed the U.S. military into Iraq with relief operations and the Good Word, but it might not be going so well. "The head of Iraq's largest Christian community denounced American evangelical missionaries in his country on Thursday for what he said were attempts to convert poor Muslims by flashing money and smart cars," says a 2004 report from Reuters. "Patriarch Emmanuel Delly, head of the Chaldean Catholic Church, told journalists that many Protestant activists had come to Iraq after the overthrow of Saddam Hussein in 2003 and set up what he called 'boutiques' to attract converts."

Such involvement mirrors the anticommunism campaigns, anti-U.N. diatribes, and pro-Israel activism that the Christian Right backed for years. But what is different today is how much more sophisticated their global presence is—religious globalization, in effect—and how much influence they've gained over

Washington's foreign policy. While the precise objectives of the many different sects are not identical, and some are frighteningly imperialistic, they share the belief that all peoples of the earth should be converted to their brand of religion, social strictures, and conservative politics. Where people need condoms and drugs for HIV or birth control, they get lectures on not having sex. Where they need broadly enforceable human rights, they get a religious bill of rights. Where they need good schools, they get churches instead. Where they need help or understanding, they get confrontation and the Gospel.

Infused with moral righteousness and what one might call "irrational exuberance," the evangelicals threaten to take some of the best practices of Americans abroad (charity and promotion of freedom) and submerge them in a tidal wave of superstition, attacks on local religions and social orders, American bravura, and outright militancy.

14   Petroleum Dependency

Everyone knows America is too reliant on petroleum—even Bush admitted, finally, that we're addicted to oil—and that too much of that petroleum comes from the Middle East and other places we consider to be unstable. Not everyone realizes just how damaging this dependency is, how many poor choices it leads us into, and why it persists.

Let's start with the basics:

- Every day, Americans pay $390 million for foreign oil, with half of every dollar going to OPEC and a quarter of every dollar to the oil-rich nations of the Persian Gulf.

- Americans collectively spend $200,000 per minute—$13 million per hour—on foreign oil, and more than $25 billion a year on Persian Gulf oil imports alone.
- OPEC countries are profiting handsomely from surging oil prices; they are expected to earn $300 billion by the end of the year.
- Unless things change, the future holds more of the same. Middle East countries hold two-thirds of the world's proven oil reserves. By 2025, the Middle East is expected to supply 36 percent of the world's oil, with OPEC as a whole producing 46 percent.
- In 2025, the United States is projected to consume 28.3 million barrels a day—44 percent more oil than we do today.

We consume so much because:

- Despite improved automotive technology, fuel economy is lower today than in the mid-1980s. In fact, in 2002, the average fuel economy of new vehicles sold was at its lowest point since 1981.
- Light trucks, including SUVs, account for 50 percent of the passenger vehicle market.
- For the 2003 model year, the fuel economy for the average car was 24.8 miles per gallon; for the average SUV and light truck, it's 17.7 miles per gallon.
- Without new technology, safe, comfortable, and affordable cars could get 50 miles per gallon or more.
- There are 200 million cars in America.
- Electric power is mainly generated by coal—horribly polluting, dangerous to mine—but oil also is used to fire about 5 percent overall.
- About 8 percent of households are heated by oil. Many more are now fueled by natural gas (electricity plants also), much of

which is being imported, often in a hazardous liquefied form. More than half of those imports come from the Middle East too.

We import so much because:

- We consume so much, and we do not produce as much oil as we once did.
- Domestic output reached its peak output of 11.6 million barrels per day in 1972. Domestic production is now about 9 million barrels per day and will continue to decline.
- We have been fuel switching from oil to natural gas; gas reserves remain steady in the United States, but this has not stopped the increasing import of oil. We have doubled our use of gasoline since 1966, for example. And gasoline accounts for half of all oil consumed.
- New production in the United States cannot make a dent in these import figures because fuel switching has already occurred, gasoline consumption steadily rises, and even the major source of new oil, the Arctic National Wildlife Refuge, would yield less than a six months' supply.
- Renewable energy consumption actually declined during the Bush administration, despite skyrocketing prices for conventional fuels.

The other costs of oil dependency include:

- Military expenditures. Even without the war in Iraq launched in 2003, the United States was spending $50 billion annually to patrol the Persian Gulf.
- Growing amounts of military attention will be paid to the Caspian and other venues of production.
- The cost of the war in Iraq is expected to be $2 trillion or more.

- The "resource curse" of sudden natural resource wealth, especially oil, in developing countries is destabilizing and often leads to conflict.
- Climate change, which is caused mainly by burning oil and coal.

The costs are tolerated because:

- The United States has no energy policy to relieve us of petroleum use.
- Oil and utility companies such as ExxonMobil and Southern spent $367 million in 2003–05 pushing Congress to pass energy legislation that was immensely favorable to their companies.
- Between 2000 and 2004, the oil and gas industry poured more than $75 million into political campaigns; four out of every five of those dollars went to Republican Party candidates.
- Renewable energy use—which could triple quickly with a minimum of cost—involves a dispersed system with many small companies providing more efficiency, conservation, and new sources of supply, and they are disadvantaged by a political system in the grip of oil corporations.
- We think we have a right to the world's oil and gas reserves.

15   The American Dream

The Dream is different things to different people. In many versions, it does capture what is good about America, particularly the notion that immigrants can make a better life here. But the Dream teeters on a fulcrum of individualism: that through pluck, hard work, and a personal relationship with God, one can become prosperous.

It is an idea that holds enormous appeal for the many millions in much of the world who have so little. But it is based on a false promise. Success in the world—financial and otherwise—does often require hard work and a certain capacity for risk taking. But gumption isn't enough.

That does not mean that one should give up on hard work and finding routes to success. But the problem with the Dream is how it locates the entire responsibility for success in this notion of individual effort. This distinctly American mythology posits a world in which it is *only* the energy and drive of the individual that matters. Good schools, healthy environs, nutritious foods, stable families, social peace, global equity, respect for diversity—fundamental qualities that allow individuals to advance, and are *common* efforts, consequences of societal decisions and enforcement—are the pillars of any person's success.

So the trumpeting of the American Dream not only creates many false expectations about how one might make it in the world, but it distorts the social and political choices necessary in many countries to provide the basics of food security, good housing, health care, and gainful employment. If the constant refrain of individualism drowns out the centuries-old emphasis on sharing, common wealth, and collective responsibility, the people of the third world particularly will be disadvantaged further by the ideology of individualism.

In one of his encyclicals, the *Centessimus Annus* (1991), Pope John Paul II put it well: "The more that individuals are defenseless within a given society, the more they require the care and concern of others, and in particular the intervention of governmental authority. In this way what we nowadays call the principle of solidarity, the validity of which both in the internal order of each nation and in the international order, is clearly seen to be one of the fundamental principles of the Christian view of social and political organization."

Solidarity is the caring of the community for all its members, and the work of the community is for all its members. John Paul refers to this, quoting Pope Paul VI, as a "civilization of love." It is not a doctrine of individualism, to be sure. That belief in the power of the well-directed individual to overcome all possible restraints is a distinctly American idea, derived, perhaps erroneously, from Emerson and Thoreau. The man is made by himself in dialogue with his Creator; he transcends the social milieu and its many constraints to achieve self-reliance.

This powerful notion of the individual at the center of the moral universe coincidentally works well for economic ideologies that seek to diminish or destroy the role of political institutions in social and economic life. So it's not surprising that individualism becomes the philosophical underpinning for American capitalism and its global reach. You downtrodden Ghanaian, Indian, Brazilian—you, too, can succeed if only your government would let us help you by lowering barriers to foreign investment, or eliminating labor laws, or breaking up and selling this inefficient state-run industry. Then the Dream is at your doorstep.

It's not quite that blatant, of course, but the pitch is similar. Underlying economic globalization is the Dream. All you need to do is work hard and be faithful. The social context does not matter. If anything, it's likely to get in your way.

In that remarkable encyclical, John Paul described how things get in the way. It's worth quoting at length, not only because he spoke as a leading conservative, but because the cause of Christian charity and fairness is a hard-won perspective of centuries. It is a coda not only for the peculiar allure of the American promised land of economic wealth, but the actual consequences of neoliberal economics in the developing world. In this passage, he was speaking of the four-fifths of humanity that is poor and have need of more than a dream.

They have no possibility of acquiring the basic knowledge which would enable them to express their creativity and develop their potential. They have no way of entering the network of knowledge and intercommunication which would enable them to see their qualities appreciated and utilized. Thus, if not actually exploited, they are to a great extent marginalized; economic development takes place over their heads, so to speak, when it does not actually reduce the already narrow scope of their old subsistence economies. They are unable to compete against the goods which are produced in ways which are new and which properly respond to needs, needs which they had previously been accustomed to meeting through traditional forms of organization. Allured by the dazzle of an opulence which is beyond their reach, and at the same time driven by necessity, these people crowd the cities of the Third World where they are often without cultural roots, and where they are exposed to situations of violent uncertainty, without the possibility of becoming integrated. Their dignity is not acknowledged in any real way. . . .

Amen.

16   The ABCs of HIV/AIDS
How Not to Stop an Epidemic

When the AIDS epidemic first became alarming in the early 1980s, treatments and even the possibility of cure in America, where it was prevalent, were considered improbable in the near term. Those cursed with the disease were stigmatized, and it took years of activism—much of it from Hollywood—and pressure from health professionals to shake the establishment into taking the problem seriously. By virtually all accounts, the first president to have to deal with it, Ronald Reagan, reacted slowly. But finally, gradually, the disease was addressed, and some palliatives, particularly antiretroviral drugs, and preventive behaviors, especially the use of condoms, began to bring some optimism to the grisly scene.

As nearly everyone now realizes, the success of the drugs in particular was confined to the global north, where they were affordable. It was not only the "patent rights" of the pharmaceutical companies that stood in the way of distribution of these life-saving drugs. It has been the policies of the Bush administration in particular, the so-called ABC guidelines, which have retarded progress in battling the disease.

And battling the disease with a major investment of resources is urgent. Many observers worry that the pandemic, tragic in the loss of life and the numbers of children orphaned, can lead to social dissolution and the collapse of societies, with further misery the result. Africa is especially vulnerable: approximately 64 percent of HIV-infected people worldwide live in sub-Saharan Africa. Currently over 5 million people in South Africa alone are infected. It is estimated that 8,500 people die every day from AIDS worldwide.

U.S. policy is flawed on several counts. First, its much-bally-hooed commitment of $15 billion, pledged to prevent 7 million new infections and treat 2 million with antiretroviral drugs (ARVs), was not all new money and has in any case not been requested or appropriated at that level. Washington insists on paying Big Pharma the inflated prices rather than negotiate for a reduction, dramatically reducing—by a factor of four—the number of patients who can be treated. Generics have been endorsed by all other parties involved in AIDS prevention and treatment, such as the World Health

Organization and the Global Fund to Fight AIDS, Tuberculosis and Malaria. The large outlays for pricey drugs recall a major criticism of U.S. foreign aid generally—that the main beneficiaries are American vendors selling services or goods.

Washington's ABC approach—Abstain, Be Faithful, or Use Condoms—has been shown to be ineffective. Several studies of abstinence and faithfulness have now been conducted, and they do not work, certainly not as the lead initiative to prevent such a devastating social phenomenon as HIV infections. One such 2005 review from Human Rights Watch is worth quoting at length:

> U.S. officials systematically ignore independent evaluations of abstinence-only programs, instead making broad and unscientific claims about the benefits of abstinence. The U.S. global AIDS strategy, for example, posits that "[d]elaying first sexual intercourse by even a year can have significant impact on the health and well-being of adolescents and on the progress of the epidemic in communities." Beyond failing to cite evidence for this claim, the strategy neglects to mention that some countries with higher average ages of sexual debut than Uganda— Zimbabwe and South Africa, for example—have much higher rates of HIV incidence. The important point is that delaying sex does not protect people from HIV unless they protect themselves once they become sexually active. Abstinence-only programs in fact increase HIV risk by withholding information about contraception and safer sex and by suggesting that married people are safe from HIV infection.

Abstinence and faithfulness became the leading edge of U.S. policy because of Christian evangelical groups in the United States and Africa. These faith-based groups can convert souls and distribute information about abstinence on the American dime—actually, on the American $1 billion or more allocated annually for abstinence programming in the fifteen African countries where the United States is active.

In fact, the policy is even more pernicious. Nearly two-thirds of all U.S. funds must go to pro-abstinence programs, and are not available to any organizations with clinics that offer abortion services or even counseling. The United States is also opposed to the provision of clean needles and syringes to drug users on the grounds that it could be construed as encouraging their habit. Similarly, condoms are restricted to the likes of sex workers and forbidden to youth, who are having sex in much larger numbers than the supposedly high-risk groups. "The U.S. has threatened to cut funding for condom provision," says one report, "stirring Stephen Lewis, the highly vocal (and Canadian) U.N. special envoy for HIV/AIDS to Africa, to proclaim this is 'a dogma-driven policy that is fundamentally flawed (and is) doing damage to Africa.'"

The multilateral approach via the Global Fund has been scorned by the United States even as the scientific evidence suggests the fund's superior approaches are more likely to succeed. There are now 40 million people worldwide who are infected, and there are deep concerns about infection rates in India and China, among other places. Where governments have dealt with the disease forthrightly, with cheap ARVs, condoms, and clean needles, as in Brazil, infection rates have been halved.

The capture of U.S. policy by evangelicals and drug companies is shameful. It is not only a waste of money, it is again sending the message that our "values" are more important than saving lives and helping poor societies cope with disease. One would have thought that the United States had evolved from the dark days of denial and condemnation in the early 1980s, but we have merely turned those sentiments into stark profit-making and insidious evangelizing.

17   Reaganism

Most Republican political leaders of government remain vocally committed to the self description of a "Reagan Republican." It's a revealing testament of sorts. For throughout his long years of campaigning for the presidency, Ronald Reagan would often say that government was not the solution to our problems, government *was* the problem. It was the perfect embodiment of his thinking about government involvement in the economy, and he applied it not only to the United States but to global economics as well, with equally unfair and unproductive consequences.

This philosophy resonated with many Americans, the vision of self reliance and rhetorical battering of city hall. Reagan was a masterful politician, and what he said and believed had the force of his winning personality and optimism. It is always easy to look at any bureaucracy, not least those of government, as too large and inefficient, and so his rudimentary notion that too much government was wasteful and in the way of the initiative of ordinary people was in some ways irresistible.

The Reagan revolution, as many of his adherents liked to call it, consisted mainly of this impulse to reduce government's size and taxes. His other impulses—to wage an aggressive Cold War and to articulate a conservative social agenda—required much larger and more intrusive government, of course, so his philosophy, if that is even the right word, was fundamentally contradictory, as indeed this brand of conservatism always was. (Those right-wing pundits who claim that free market policies and security policies can be separate realms are ignoring the plain fact that security involves an all-embracing economic policy, from massive subsidies to the defense industry and acceptance of those business models to the use of military power to enforce economic policy abroad. They are inextricable.)

The Cold War agenda was very much in keeping with the long-standing policies and values of successive presidencies, a rather smooth progression of military readiness, anticommunism, and alliance building. (Read Kennedy's inaugural address for a really belligerent piece of Cold War rhetoric.) The Reagan Doctrine was itself a continuation of sorts, though more intense in its catastrophic consequences. Reagan showed remarkable nimbleness in reversing himself on his anti-Soviet posture when it suited him, after the Iran-contra scandal, a bit of politicking that his base never quite understood. The social agenda was pretty much a run-of-the-mill right-wing screed against crime, permissiveness, Mexicans illicitly entering the country, liberal judges, the ACLU, and abortion—a sturdy set of distractions if there ever was one.

It was instead this notion of less government where he had a more lasting impact. The claim that he somehow ushered in a remarkable new era of prosperity and private initiative in America is part of the phony legend. The percentage of GNP that goes to government spending was 21.6 percent at the beginning of his presidency in 1981, at 21.8 percent at the end of his presidency in 1989. (Of course, the funds shifted from human services to the military and debt service, the latter shooting up by 70 percent.) The growth in the economy was built on debt, the same way a family can look prosperous by filling up several credit cards. Even then, inequality increased, personal income for all but the well off stagnated, the U.S. trade deficit soared, and personal and state government debt skyrocketed, too. Health and safety regulations, and those to protect the environment, were under constant assault, as were labor unions. This was the corporate agenda—a shift of resources to military contracts, lower taxes for the executive and leisure class, deregulation, "defunding" the left—that formed after the tepid government activism of the previous fifteen years.

Applying this to the global economy was the big game, however, and here Reagan succeeded perhaps beyond expectations. Reagan's pursuit of trade agreements, World Bank and IMF mandates, and foreign aid and loan regulations—all, as it happens, using the leverage of the U.S. government—built in a set of conditions and momentum that have been difficult if not impossible to reverse, despite the evidence of poor results. The "magic" of the marketplace and less government and high military spending just do not soar in most places, and most places don't have the big cushions we do to break their fall. One consequence has been a great deal of misery, perhaps unprecedented in scale, in the developing world.

This, of course, is not the Reaganism of yore, which has the tall cowboy riding in on his steed to glare down the Russkies and make America proud again. The legend does not bear close scrutiny, as several of our *100 Ways* attest. Reaganism—the attack on government as a source of income stability, job creation, protection of the weak against the predatory strong—does live on, however, in the current administration and in imitators around the world, and the results are almost always the same: the wealthy do well, the poor and middle classes do not. The natural environment is trashed, the military sucks up more of the national treasure, immigrants are demonized. The most tragic legacy of Reaganism, though, is the proliferation of the antigovernment mantra. We need to think creatively about how governance occurs, how it is accountable, whom it serves, what its size and proper functions are. But Reaganism promoted none of that fresh thinking. His was a simple (he was very very good at simple) gimmick: blame government, but use it as much as possible for your own ends.

That kind of cynicism rules the right wing and has since 1980. His successor, a conservative corporatist, George H. W. Bush, promised the nation "a kinder, gentler" administration and made this a major theme of his White House campaign—Reagan's own vice president essentially embarrassed by Reaganism's harshness. He was never forgiven by the base. The clumsy apotheosis, of course, is George W. Bush, who has used the power of the government in extraordinary ways to make war (even Reagan did nothing as stupid as invade Iraq; of course, he cozied up to Saddam rather easily), snoop, torture, push the social reactionaries' agenda, and exploit nature as never before. The awkward moment of George W. reaching for the mantle of Reagan at the Gipper's funeral was pitiful but telling. At least Reagan was a self-made man.

But that quality, part real and part myth, was perhaps what betrayed his better angels—the notion that everyone can be self-made the way he was in his own Hollywood style. Whatever its origins, it turned into an ugly ideology, the revolt of the rich. And most sadly, it affected not just America but spread its ill effects throughout the world. It is a poisonous legacy, sunny optimism and all.

18   Nuclear Weapons

When the first two atomic bombs destroyed the Japanese cities of Hirsohima and Nagasaki in 1945, the relief that the Second World War was ending trumped concerns about the new weapon. (The relief might not have been felt by the residents of those two cities, with their 120,000 dead.) American ingenuity had turned the secrets of physics into an astonishing and frightening agent of global war. It could have ended there. But it did not, and we have paid a very high price ever since.

One can fault the development of the "ultimate weapon," but it was pursued to defeat Nazism and hardly anyone would object at the time. That it was employed against a Japan that was close to surrender was probably done to impress the Soviets as well as the Japanese, an interesting act of what today would be regarded as terrorism. But whatever the politics of the only use of nuclear weapons in warfare in history, it was possible to discourage development of nuclear capability right then and there.

Some diplomatic attempts to do just that faltered. Historians blame not just Stalin, but Truman and a conservative Congress that regarded nuclear weapons as the harbinger and guarantor of American global power. The weakened attempt to prevent the oncoming nuclear era, mainly in the plan of financier-turned-diplomat Bernard Baruch, might never have worked, given Stalin's nuclear aspirations, to say nothing of France and Britain. But Baruch's ideas were never given a chance.

The results of that failure, despite formal treaty obligations to disarm—still ignored, even after the collapse of the Soviet Union—and a worldwide aversion to nuclear weapons, have been wasteful, poisonous, and potentially now a new threat to the United States and others via nuclear terrorism. The nuclear arsenals of the United States grew to 34,000 strategic and tactical weapons. The cost of this, including the technology needed to deliver weapons, has been $5.7 trillion so far. That's $5,700,000,000,000. The Soviet nuclear arsenal reached a similar size.

But the costs are not calculable only in dollars. The mines where uranium is extracted from earth, the industrial factories where nuclear materials like plutonium and enriched uranium are manufactured, and other facilities where nuclear weapons are fabricated, also produced nuclear pollution and unsafe conditions for the workers in this vast nuclear complex. They and those around this complex suffered what one former assistant secretary of energy called "slow motion nuclear war." Small in scale compared with the death tolls in Japan, perhaps, but the effects of this nuclear fuel cycle are poisonous all the same. What happened in similar communities in the Soviet Union is equally horrific, if not worse.

In the early years of the arms race, nuclear weapons were tested in the atmosphere, and this also caused disease downwind of the test sites until President Kennedy called for and achieved an atmospheric nuclear test ban in 1963. (The Test Ban Treaty was opposed by the Joint Chiefs of Staff and many conservatives at the time, insisting the treaty would fatally weaken the United States.)

Some argue that nuclear weapons kept the peace between the United States and the Soviet Union during the Cold War. There was peace between the Soviet Union and the United States for the nearly thirty years before the invention of nuclear weapons, however, and the Cold War itself may have been a product of the fear that possession of nuclear weapons sparked. But however one argues this point, it is certainly true that the *scale* of the arsenals, the arms race, and the resultant costs of that race were wholly unnecessary. Fifty or so weapons would have deterred Stalin as much as the thousands that eventually were deployed. (And Stalin was dead by 1954, very early in the nuclear era, and even he was deterred from further mischief at a time when there were very low numbers of nuclear weapons.)

Fortunately, public concern about nuclear weapons throughout the world forced political elites to pursue an arms control and disarmament agenda, however imperfectly. The 1963 accord was the first of several important agreements, and the 1968 Nuclear Non-Proliferation Treaty set up a mechanism to prevent civilian nuclear power activities from being a springboard to nuclear weapons. It has worked somewhat well. Many useful mechanisms for reducing the threat of nuclear explosions have been fumbled away, including a ban on the production of the fissile materials needed for bombs. It has been sustained pressure from peace movements, however, that achieved the breakthroughs in reducing the most perilous confrontations between the United States and the Soviet Union.

Today, nuclear weapons remain a threat, though of a different kind. Nuclear arms are now possessed by China, India, Pakistan, and Israel, as well as Russia, Britain, France, and the United States. Others have wanted to develop them, notably Saddam's Iraq, Qaddafi's Libya, and the apartheid regime in South Africa—

though they all were very far from any capability—and now possibly by Iran. More worrisome than the dark fantasies of nuclear weapons in the quiver of tinpot dictators is the potential for al Qaeda or other nihilist groups obtaining a weapon or two and exploding one in Tel Aviv or Tokyo or New York. While this is also extremely difficult to do—if states like Iraq with oil money and scientists and easy access to black markets cannot do it, an Osama bin Laden or Tim McVeigh are far less likely—it remains the direct legacy of the nuclear era begun more than six decades ago in Los Alamos, New Mexico, and announced to the world in the deadly explosions in Hiroshima and Nagasaki.

19   Genocide

"An individual death is a tragedy, a million deaths are a statistic," Stalin once said. When it comes to millions of deaths, Stalin knew whereof he spoke. But while he and Hitler and Mao and Enver Pasha and numbers of other murderous despots litter the genocide hall of shame, the United States has its own uneasy relationship to mass killings of innocent people. It's not a polite topic, of course, and it's not a simple one. But throughout its history, the United States of America has much to answer for when it comes to large-scale massacres.

Almost all genocides or massacres are justified by resort to a warrior logic of one kind or another. Stalin's was collectivization and counterrevolutionaries. Enver Pasha, the prime minister of Ottoman Turkey when it killed 600,000 Armenians during the First World War, saw the Armenians as allies of the enemy Russia and a national security threat. Mao also had his enemies-of-the-people excuses. Hitler was more obviously a psychopath, but nonetheless created the German Aryan myths and doctrines of subhuman peoples, namely, the Jews and Slavs.

The "not human" notion is salient to our own particular shames. The white settlers of the continent eradicated 85 percent of all indigenous tribes and perhaps 8 million to 10 million people, a slow motion genocide but a very thorough one. The excuse was that they weren't really fully human, and they were in the way. Just about everybody joined in, from national heroes like Andy Jackson and his slaughtering of the Seminoles, to virtually every military operation as European settlers moved west. The U.S. cavalry, not the settlers themselves, did most of the murdering, and the native tribes never really had a chance when serious firepower was applied.

After the Battle of Wounded Knee in South Dakota in 1890, the game was up for the victims, and they gradually were forced into concentration camps—excuse me, "reservations"—and descended into poverty, alcoholism, and despair. (I always found it piquant that the successor to the cavalry, the army's helicopter forces, adopted names for their whirlybirds like Commanche, Black Hawk, Kiowa Warrior, etc., as if killing off these tribes was somehow honored by stealing their names.) Wounded Knee, a mere twenty-five years before the Armenian slaughter in Anatolia and forty-eight years before the Nazis' Kristallnacht, is rarely mentioned when the genocides in Bosnia, Rwanda, or Sudan are fretted over on the op-ed pages. Historical significance apparently has a statute of limitations.

The U.S. military moved on to new exterminations elsewhere. The slaughter in the Philippines in the early twentieth century is a grisly tale, and, remarkably, not much taught to our school-children. From 1898 to 1905, U.S. forces massacred 500,000 or more Filipinos, a small fraction of whom were rebelling for their sovereign rights. These were not casualties of war in most cases; they were savage and racist mass murders. Some nasty interventions in Central America probably do not qualify as genocides, but in small countries even a little killing goes a long way. The atomic bombings of Hiroshima and Nagasaki killed more than 100,000 civilians in two radioactive blasts, unnecessary and unwarranted.

Regrettably, it doesn't end there. We have sins of commission, and sins of omission. The latter include not speaking up when we know something is amiss, or not acting to prevent the mayhem when we have the means to do so. Many historians believe that FDR was slow to react to knowledge of the Nazi death camps. Our deposing of democratic government in Guatemala resulted in tens of thousands of indigenous peoples being killed by the military dictators the United States subsequently supported. (Ríos Montt, the worst of them, was publicly embraced and celebrated by Ronald Reagan.) U.S. presidents have long been willing to turn a blind eye to massacres if committed by useful allies, however bloodthirsty. The list is quite long—the nightmares for Kurds in Turkey and Kurds and Shia in Iraq, East Timorese, the Sudanese in Darfur, and Cambodia. In some cases, the crimes were perpetrated with American weapons, like the Turkish army's ethnic cleansing of Kurds; in other cases, it was a consequence of misbegotten policies, like the U.S. debacle in Southeast Asia and the Cambodian genocide that followed in its wake.

Possibly the most notorious turning away from genocide had little to do with realpolitik and everything to do with cowardice— the willful ignorance of the Clinton administration on the eve of, and in the midst of, the Rwandan genocide in 1994. Between 500,000 and one million people died in the span of a few weeks. The United States, knowing full well what was happening, sat on its hands. The U.N. ambassador, Madeleine Albright, later to become secretary of state, played an especially shameful role in the spinelessness. That this was taking place while the slaughter of Bosnians was also proceeding is all the more remarkable.

Again, one can say that big bold countries make mistakes, and that ours pale before the crimes of the Russians and Chinese and even our pals the Brits. But actions have reactions, and the consequences of bad behavior, sometimes repeated or unacknowledged, is to give license to others. No less a genocidal psychotic than Osama bin Laden has cited our terrorism in Hiroshima as a riposte to the many declamations of his activities. But one needn't go to the mountaintop to find such sentiments. The "street," whether Arab or Latin American or Indochinese, regards our moral posturing through a lens of bitter memory.

The Convention on Genocide, a U.N. treaty that has been signed and ratified by all civilized countries, would impose penalties on the guilty and bring needed attention to the crimes themselves. The United States has not acceded to all its provisions. Now there's a message to take to the rest of the world.

20–27   A Rogues' Gallery of Dictators

When the Shah of Iran left Tehran for good in 1978 and the Ayatollah Khomeini was on his way to power, I had a conversation with a colleague at a newsmagazine, a conservative fellow who I thought would be devastated by the news. "Well," he said with a smile, "he was our guy and we had him there for twenty-five years. Not bad."

"Our guy" was a linchpin of Cold War thinking. Lots of "our guy" governments—repressive, nondemocratic, corrupt, sometimes genocidal—dotted the globe, in power mainly at the pleasure of Washington. Some were here and gone quickly. Some fit certain patterns of oligarchs, military strongmen, business elites. Whatever their exact nature or origins, they all play a set-piece role in the Cold War drama scripted in Washington: General So-and-So has stepped in to restore order after unrest caused by provocateurs, and he will run the government like a business, end corruption, strengthen the fight against communism, and soon restore democracy with new elections. An assistant secretary of state would be dispatched to have frank discussions with the general, at which time the U.S. government was assured of the new leader's good intentions. An aid and trade package (with foreign investment included) would be signed as well. Moscow, of course, had its own version of this, but we were much more adept at the play, could put more people in the seats, and built the theater.

The pattern was so predictable that we can abbreviate the next eight Ways by brief portraits of some of my favorites. (Others, like the shah or the Duvaliers, are detailed elsewhere.) Almost all are typical examples of U.S. Cold War policy.

20   Augusto Pinochet and Chile

Richard Nixon and Henry Kissinger could not abide leftist leaders in Latin America, "our backyard," and none was so challenging as the democratically elected Salvador Allende, a socialist who set out in 1970 to nationalize Chile's industries, including some owned by Americans. A potentially successful democratic Marxist in a large country in the Western Hemisphere was unacceptable, so Nixon and Kissinger unseated Allende in a CIA-backed coup and installed General Pinochet in 1973.

Pinochet's seventeen-year rule was with the proverbial iron fist. More than 3,000 Chileans were executed or "disappeared" for being liberals or leftists. Some 30,000 were incarcerated, with many of those tortured, and another 250,000 were detained. Pinochet had Allende's foreign minister, Orlando Letelier, assassinated in Washington in 1978, with no repercussions. Free market economists of the "Chicago School" were imported to reform Chile's economy, and their experiment failed to address the inequalities plaguing most Latin countries: poverty doubled during Pinochet's dictatorship, with nearly half of all citizens living in poverty by the end of his reign of terror.

Allende's overthrow and death, and Pinochet's ruthlessness, were intended as a lesson to all leftists in Latin America: try to promote equitable social policies and you will be crushed. This was Chile's "9/11," and the repercussions are still vibrating.

21   The Argentine Generals and the "Dirty War"

The southern cone had other notorious dictators, among them Stroessner of Paraguay and military cliques in Uruguay and in neighboring Brazil. But perhaps the bloodiest period of repression was the reign of the military junta in Argentina from 1976 to 1982. Built on the repressive and increasingly conservative reign of the Peróns, the generals—Valdera, Viola, and Galtieri—ran a state terror operation of breathtaking brutality. Thirty thousand Argentines disappeared. Thousands were jailed and tortured.

Their reign was economically as well as morally ruinous, but the Reagan administration did its best to prop them up. Jeanne Kirkpatrick, Reagan's U.N. ambassador, provided a cloak of respectability by visiting Buenos Aires in 1981, but she really showed what Reagan & Co. thought of human rights and democracy by snubbing the mothers of the disappeared, a respected human rights group, who asked to meet with her.

A year later, facing growing internal opposition to its rule, the generals launched the ill-fated military campaign against the British-held Falkland Islands. It was Kirkpatrick again who stunned the world by initially siding with Argentina, a bloody dictatorship, over Britain, our longtime democratic ally. But even the Reaganites could not save the generals of Argentina.

22   Mubarak and Egypt

The world's most populous Arab country provides a different kind of American involvement in a repressive regime. Egypt's history is complex, fraught with foreign intrigue for hundreds of years. The United States was long wary of Egypt's intentions during the early Cold War, when it was led by the charismatic nationalist General Gamal Abdul Nasser, who gradually accepted Soviet aid and advisers. After Anwar Sadat succeeded to the presidency after

Nasser's death in 1970, Egypt ousted the Soviets but maintained antagonism toward Israel, launching a war in 1973. Sadat ultimately made peace with Israel, a major achievement, when Jimmy Carter was president.

Just after Reagan was elected, Sadat cracked down on internal opponents and was assassinated for it in 1981. His vice president, Hosni Mubarak, a former general, came to power and has remained ever since, bolstered by massive U.S. assistance and an attitude of looking the other way when Mubarak prohibits free elections and ravages his opposition. His utility in the grander American scheme after the Cold War ended has been as a bulwark against Islamic extremism and as a partner with Israel in the Middle East peace process.

But Mubarak, along with the Saudi royal family and other repressive rulers in the region, is now a symbol of American acquiescence and support of repressive Arab regimes, and this is one of the most attractive (and undeniable) assertions of the jihadists.

23   Pol Pot and the Cambodian Genocide

After the U.S. debacle in Vietnam, all of Indochina was a mess. The biggest mess of all was Cambodia, where the U.S. bombers and troop incursions, covert action and political intrigue, doomed any prospect of stability during the long American war against the Vietnamese. The Khmer Rouge, a Marxist guerrilla army led by the murderous Pol Pot, seized on the instability and violence to take power by force in 1975. Estimates of the genocide he committed range widely; up to one million Cambodians died. Millions more were refugees. Cambodian society, already devastated by U.S. interventions—itself a reason why the Khmer Rouge could temporarily triumph—was laid to ruin.

Bleeding from the neighbor's woes, Vietnam moved into Cambodia and installed a friendly regime in 1979. America stood with Pol Pot. Starting in the Carter White House and continuing throughout the Reagan years, the policy of backing Pol Pot was achieved with political support, direct financial aid ($85 million or more), and encouraging China to supply him even more lavishly. Another decade of war and instability prevailed until the Vietnamese withdrew. Only a heroic effort by the United Nations has brought Cambodia partially back to becoming a livable place.

The journalist William Shawcross perhaps summarizes it best: "Neither the United States nor its friends nor those who are caught helplessly in its embrace are well served when its leaders act, as Nixon and Kissinger acted, without care. Cambodia was not a mistake; it was a crime." Add in Carter and Reagan's support for Pol Pot, and another set of crimes is apparent, with millions suffering as a result.

24   Suharto and Indonesia

Concern about the communist advances in Asia obsessed the U.S. government for decades and led to many unsavory episodes, Vietnam and Cambodia being just two among them. We supported strongmen in South Korea and the Philippines until popular movements forced them out. One of the grislier stories can be found in Indonesia, the world's fourth most populous country.

For twenty years after the Second World War, Indonesia was run by Sukarno, a populist and nationalist, who was finally ousted in a coup in 1965. His successor was Suharto, a military man and avowed anticommunist. The circumstances of Suharto's ascension remain cloaked in some mystery. The "year of living dangerously" included an assassination of anticommunist generals as a pretext for seizing power, and a subsequent massacre of 500,000 citizens—alleged communists—by the government under Suharto's emergency rule. It is widely believed that U.S. agents were involved in the massacre, encouraging Suharto and supplying names of victims.

The dictatorship Suharto established with Washington's complicity lasted for more than thirty years, and included suppression of East Timor, Aceh, and other Indonesian possessions, often with a brutality that aroused the world. Most important, the destruction of the left and the human rights abuses of the regime may have given rise to Indonesia's Islamic militancy—it is the most populous Muslim country in the world.

Suharto left office, as a result of the Asian financial crisis, with a personal fortune of $15 billion.

25   Mohammad Zia ul-Haq and the Militarization of Pakistan

One of the seemingly good legacies of British imperialism was the durability of democratic institutions. The South Asian subcontinent was for a time Exhibit A, but Pakistan's devolution into militarism has scotched that idea. No one was more responsible than General Zia, who came to power in 1977 in a coup against the elected Zulfikar Ali Bhutto, whom he then hanged. Zia was U.S.-trained, and like all Pakistani strongmen, maintained close ties to Washington, which "balanced" India's independent streak in this way.

Zia crushed most of Pakistan's democratic traditions, enlarging the role of the executive, and even introduced "Islamization" into high-level Pakistani politics. He also ensured that the military would always have a special place in the country's governance, which it certainly has. Pakistani intelligence services have been linked to terrorism in India, terrorism and insurgency in Kashmir, and who knows what in Afghanistan.

The latter country has a special place in the Zia story, because the Soviet occupation strengthened Zia as he provided a platform for U.S. operations in support of the mujahideen. Internal repression, the development of nuclear weapons, the fomenting of jihad-ist insurgencies—these actions paled, in Reagan's eyes, before the all important goal of bolstering the Islamic militancy of the muj, and Zia was the happy recipient of lavish U.S. support. His mysterious death in a plane crash in 1988 did not affect U.S. support of Pakistan's military, which continues to this day.

26   Mobutu Sese Seko, America's Congolese Cold Warrior

His name means "the all-powerful warrior who, because of his endurance and inflexible will to win, will go from conquest to conquest leaving fire in his wake," and he came to power in this populous, resource-rich, and ethnically diverse central African nation with the connivance of the CIA and the U.S.-condoned murder of the charismatic leftist leader Patrice Lumumba. Mobutu Sese Seko, a general, ruled from 1965 until unrest fomented by the genocide in Rwanda stirred a militant opposition movement against him in the mid-1990s. Thirty years of U.S.-backed rule left the country in shambles, and the consequent civil war left three million dead and the country a cesspool of instability.

Mobutu was a paragon of a kleptocrat and ruthless dictator, executing opponents at will and stealing $5 billion for his Swiss bank accounts. The United States nonetheless continued to support him for the usual Cold War reasons.

Consider, for example, President Reagan's speech after meeting with Mobutu at the White House: "President Mobutu and I have had the opportunity to review and renew one of our oldest and most solid friendships in Africa, that between the United States and the Republic of Zaire. Cooperation between the United States and Zaire under President Mobutu's leadership stretches back through twenty years and five U.S. administrations. In that time, American leaders have learned to place a particularly high value on President Mobutu's insights and counsel."

27   Saddam Hussein, the Durable American Friend

The pictures of Reagan envoy Donald Rumsfeld, Senators Bob Dole and Alan Simpson, and various other U.S. officials glad handing Saddam in the good ol' days when he was our guy (almost) in the Gulf and the bulwark against Khomeini remind us of how screwy U.S. foreign policy has been in the region. Saddam started out in the service of the CIA and then became a Soviet ally—most Arab socialists were in the 1960s and 1970s—but gradually came back to our side when opportunity knocked. The opportunity was America's lavish support of the shah of Iran, which collapsed on itself and brought forth the Islamic revolution. Then Saddam was Washington's dreamboat.

Reagan provided Saddam with everything but weapons (as far as we know). He was given $5 billion in credits, political credibility, and creditable "real time" satellite intelligence during the Iran-Iraq war that saved his regime. The United States blinded Iran's radar during a key battle in 1988. Saddam got militarily useful equipment like trucks and computers, some of which could be used in his early if unsuccessful nuclear weapons program. And he got the U.S. government to look the other way, essentially, when he used chemical weapons against Iran and then the Kurds. That was a lot of support, and it's not surprising that he thought Washington would go for more when he took Kuwait.

The first to cry foul against Saddam was Amnesty International, not the Republican National Committee.

But the story gets worse. Saddam was a CIA-paid agent in the late 1950s and early 1960s, part of a long-standing plot to murder the Iraqi strongman Abd al-Karim Qasim. The relationship lasted for years, and was doubtlessly maintained right up through his bloody rise to power.

He was one of our guys, in power for more than thirty years. Not bad.

28   SUVs

The secret about SUVs in the United States has been out for some years now. Gas guzzling, difficult to drive, dangerous to other motorists and pedestrians, and generally a symbol of American arrogance and stupidity, the so-called sports utility vehicles, neither sporty nor utilitarian, are socially and ecologically a disaster. Now we're taking all those fine qualities and selling them to the rest of the world.

The polluting monsters exist mainly because they are highly profitable, the salvation of the otherwise moribund American automakers. "Let's take Ford, for example," explains an auto-industry analyst. "The industry says that Ford is probably making $5,000 on each Explorer sold. So then you have an Expedition that sells for, let's say, $30,000 to $40,000. They're making $10,000 a unit on that. And they may be making as much as $20,000 a unit on a Navigator." Real money, he concludes. So the Bush administration has done its best to provide tax breaks, not enforce fuel efficiency standards, and generally promote the monsters as much as possible, and Democrats largely comply.

And now they are invading the small-is-beautiful environs of Europe and other implausible markets.

"SUV sales accounted for 5 percent of Europe's 16.5 million new-car sales in 2003," reported the *Wall Street Journal*, "up from less than 2 percent in 1990. Automotive research firm R. L. Polk

Europe expects sales to rise 46 percent to 1.2 million units during the next four years." Sales of SUVs in Europe rose by 15 percent from mid-2004 to mid-2005, while demand for ordinary cars declined by 4 percent in the same period.

The numbers in the United Kingdom are surging, and the wily Brits now have a way of coping with the social opprobrium properly heaped on SUV owners: they're selling mud to spray onto the monsters. "For 8 pounds (about $14.50), buyers get 0.75 liters (.85 quarts) of genuine filthy water, bottled from hills near . . . the rural England-Wales border. The aim, says the [company's] Web site, is 'to give your neighbors the impression you've just come back from a day's shooting or fishing—anything but driving around town all day or visiting the retail park.'"

And lest you think that Eurobrands are stealing all of America's SUV thunder, consider this gush in a London report aimed at women: "The dramatically styled Cadillac CTS, a true luxury sports saloon with a European driving feel, is spearheading Cadillac's commitment to become a global premium brand. The CTS is complemented by the versatile SRX 4x4 and by the XLR coupe/ convertible." American brands lag behind Asians, but the concept is strong and even some Detroit makers have gained in various parts of the world, like the Persian Gulf.

Europeans and Asians are perfectly capable of fouling up their own environments, but it's inconceivable that the SUV craze would have gotten traction without America's infatuation with these grotesqueries. And even if they do not spread to the rest of the world, the damage done to the environment and the intensification of our oil dependency are reasons alone to finger SUVs as a lethal hazard to life on this planet. We should at least provide them with an apt moniker—SUVs, Socioecologically Uncool Vehicles.

29   The "War on Terrorism"

When President George W. Bush declared a "war on terrorism," few Americans were reluctant to take on the detestable al Qaeda after the atrocities of September 11, 2001. The mangled syntax aside, the war on terrorism seemed like a necessary task, though its scope and methods were far from clear.

What we know now is that the WoT has not been especially successful, and where it has, it's because of old fashioned police work and prosecutions rather than anything much resembling a war. Where it has been an actual war—in Afghanistan and Iraq—it has achieved antiterrorism objectives in a very limited sense, and possibly not at all. The war in Afghanistan, which did not set out to be an exercise in regime change and nation building, changed the regime (mainly due to the Afghans themselves) from the loathsome Taliban, but the longer-term outcome, and the nation-building effort, remains uncertain. Heroin production, warlordism, and various forms of human abuse seem to be returning. Hunting down and destroying al Qaeda, and capturing or killing Osama bin Laden, has been a notable failure. Disruption, yes; destruction, far from it.

Iraq has been a near catastrophe in most ways but for the removal of Saddam, which is an achievement soaked in blood, past, present, and future. The point here is obvious to all but Fox News viewers, that there was no connection between Saddam and al Qaeda except mutual hatred.

Elsewhere, the WoT has been an exercise in futility, costliness, and human rights violations, all for very minor gains. The achievement is a very gradual roundup of jihadists in Europe and some elsewhere, Muslim men of indeterminate dangerousness. That they have been identified, arrested, and dealt with by more or less normal means of law enforcement is the key point: no war here. Even the law enforcement effort has been dodgy at times, as with the murder by police of a Brazilian man in the London subway. Civil liberties are increasingly snubbed in Europe and very readily in the former Soviet Union, parts of the Middle East, and other places where petty despots—with the enthusiastic support of Washington—use the convenience of WoT funding, intelligence operations, and political blindness to crack down on opposition groups.

At Guantánamo Bay, where the U.S. military has held hundreds of captured soldiers from Afghanistan; in the retention of suspects picked up off the streets of many foreign countries and whisked off in chains to secret prisons; in Abu Ghraib and goodness knows what other places of torture and humiliation, the U.S. government has permanently sullied America's name in a misbegotten antiterrorism crusade.

In the United States, the effort to root out a nonexistent Islamic threat has led to thousands of detentions, "special registration" for Muslim men, 300-plus prosecutions on spurious charges, the near destruction of Muslim charities and other social institutions, high bars to students and other legitimate immigrants from Muslim countries, and so on. This antiterrorism campaign has netted very, very little, has branded a large and growing American community as subversives, and has sent an ugly message around the world about our attitude toward Islam and its people (a message that tracks perfectly with the wars of choice in the Middle East). Even the 9/11 Commission Report found no domestic terrorism presence of any kind, before or after 9/11, and all the detentions and prosecutions have failed to uncover any serious threats, either.

The sum of this costly war—now reaching into the trillions of dollars—is an image of America at war with itself, Islam, its oldest allies, and the principles of international law that govern our world. The political use of the "terrorist threat" is blatant and obvious to all. That there are some who would do civilized people harm is also apparent, and should be dealt with as the Europeans are doing, rather well, thank you, and not through a hideously polarizing and expensive war that will have no end, and is producing more enemies than friends.

There is a poignant passage in the play about Thomas More, *A Man for All Seasons*, written by Robert Bolt, that is germane to the war on terrorism. Thomas More was a Catholic statesman who was a defender of political conscience during the tumult of Henry VIII's reign in England. In Bolt's play, More is defying Henry's capricious moves to reshape the law and customs of the country. He is speaking to Roper, a onetime protégé who is siding with the Crown.

Roper: That man's bad!

More: There's no law against that.

Roper: God's law!

More: Then God can arrest him.

Roper: While you talk, he's gone.

More: Go he should, if he were the Devil, until he broke the law.

Roper: Now you give the Devil benefit of law!

More: Yes, what would you do? Cut a road through the law to get after the Devil?

Roper: Yes. I'd cut down every law in England to do that.

More: And when the last law was down, and the Devil turned on you . . . where would you hide, Roper, the laws all being flat? This country is planted with laws from coast to coast . . . Man's laws, not God's, and if you cut them down . . . and you're just the man to do it . . . do you really think you could stand upright in the wind that would blow then? . . . I give the Devil benefit of law for my own safety's sake.

We don't know how long the war on terrorism will sully the United States or keep innocent people locked up or provide excuses for domestic spying. At present, it is done more for political expediency than to protect the American people. It has served Bush well, getting him reelected in 2004 and keeping Republicans a majority in Congress. The question is, will the war on terrorism remain a permanent fixture of U.S. politics and the federal bureaucracy? If it does, and chances are it will, then the harm done will be vast indeed. The law's protection cut down, here and abroad.

30–35   How to Really Screw Things Up
Six Splendid Little Wars

Any self-respecting analyst of how America has screwed up the world would dedicate far more space and verbiage to the many military interventions the U.S. government has launched, nearly always with bloody effect or bad result, sometimes both. But the tale is so commonplace, following a fateful script, that we can dispense with them in shorter form. They were launched to smite the commies (where none existed), to right a wrong (usually with American fingerprints all over it), or to burnish the reputation of a president whose standing may have been cut low by other misdeeds. He could always rely on the rough 'n' ready public to go along with a quick and decisive bit of bloodletting.

Some are more direct than others, some deadlier, some longer in gestation, all with nasty aftertastes and long festering wounds. The more significant blunders, like Vietnam and the recent invasion of Iraq, deserve and get longer treatments. These short takes will nonetheless convey the message.

30   Guatemala, 1954

The CIA's overthrow of the democratically elected Jacobo Arbenz in 1954 was not the first time the United States intervened in Latin America, of course, as a long history of military invasions, occupations, and covert actions plagued Mexico, Nicaragua, Panama, Cuba, Haiti, and many others over the years. But there was something truly evil about America's meddling with Guatemala, one of the most beautiful countries in the world, with a rich culture of indigenous peoples.

Guatemala followed the contours of many countries in the region, with great disparities of wealth and political status, dominated by an elite of European businessmen and military cliques. Indigenous peoples, descendants of the legendary Mayans, were treated shabbily, sometimes in slavery-like conditions that violated all standards of human rights. But Guatemala managed to move forward, sooner than many of its neighbors, toward democratic government and liberal reform. Arbenz was the second consecutive president to be popularly elected, and he instituted a land reform program in 1952. That was his undoing.

Land reform was a volatile topic in Latin America and the Caribbean. Because land was the major natural resource, it was highly valued and mainly in the hands of the oligarchy. Reform meant redistribution, and that was considered radical, even communistic. When U.S. companies were involved, as was the case in Guatemala—the United Fruit Company—the pressure to stop reform became quite intense. The CIA engineered a coup (they had contemplated assassinations) by fixing the region-wide political conditions, mounting the rogue army, funneling money, broadcasting propaganda, and providing air support for the seizure of power. No U.S. troops landed in Guatemala, but it was a proxy intervention of the most direct sort.

What made this episode especially odious is what occurred

over the following decades. Not only had the United States reversed a promising experiment in democracy and social justice, but it drove the point home by supporting a series of brutal dictatorships that over these ensuing years murdered tens of thousands of peasants and political activists. The worst of these dictators was Ríos Montt, whom Ronald Reagan described as "a man of great personal integrity" who is "getting a bum rap on human rights." A born-again Christian, Montt escaped justice for the 70,000 deaths of his regime. The country, unable to stabilize since 1954, has been racked by small-scale insurgencies, crime and corruption, and the continuing impoverishment of its enormous indigenous population. Had Washington allowed Arbenz to prevail, the bloodshed and misery may well have been avoided altogether.

31   Dominican Republic, 1965

Why do so many of these interventions happen in El Carib? Some of it is our notion of the warm waters to the south being our own lake, but since the late 1950s, it probably has more to do with the man with the cigar, who by 1965 had been in power in Havana for six years and was driving the third president of his tenure crazy. Lyndon Johnson, leading America into the complete disaster that was the Vietnam War, sent marines to Santo Domingo in 1965 to "protect American lives" and to counter what was described as a communist insurgency. In fact, the instability in the country, after years of a U.S. backed dictator (Trujillo), was in part the result of the United States trying to block the restoration of the democratically elected, and slightly progressive, Juan Bosch.

So the marines and the Eighty-second Airborne invaded and occupied DR, and in the invasion some 3,000 Dominicans lost their lives. It is now clear that the "communist threat" was grossly overstated, and that the lives of foreign nationals in the country—including Americans—were not in jeopardy. As a major scholar of the region puts it, "The U.S. government's preoccupation with avoiding a 'second Cuba' had structured the way American officials looked at the Dominican Republic throughout the early 1960s."

But Bosch, like Arbenz and others of the reformist bent, was not a communist and was not connected in any way to the Soviet Union or Cuba. The Dominican Republic invasion and occupation—with 23,000 soldiers—is an example of a paranoid fantasy that set back ordinary reform and ushered in a decades-long period of mismanagement and oppression. A Trujillo puppet, Joaquín Balaguer, was promoted and ruled autocratically for thirty-five years. The country remained shockingly poor throughout the period.

32   Grenada, 1983

President Reagan was searching for something—a nice quick thrashing of a commie somewhere to match his apocalyptic rhetoric. Grenada, an island in the Caribbean few of us had heard of before that fateful few days in 1983, served perfectly. It resembled Cuba in some obvious ways. It had a quasi socialist at the helm, Maurice Bishop, who had been rather successful, and a friend to Castro. Things got a little disorderly when a rival faction imprisoned him (finally executing him), which caused some turbulence, a pretext for Reagan to invade to "save American lives." But most important, Grenada was available to distract America from the catastrophic stationing of U.S. troops in Lebanon, where, just two days before U.S. troops carried out Operation Urgent Fury, 241 of their brethren were blown up by a suicide bomber in Beirut. Bad press? Change the subject.

The invasion's pretext was that some American students were in jeopardy on the island. This was bogus, of course, but it was good enough for all concerned. The entire thing was over in a matter of hours, as one would expect. A "lesson" was delivered to Fidel Castro, which he somehow never absorbed. Several dozens were killed. More medals were delivered to the participating troops for bravery than there were troops. Everybody was a winner, except, of course, Grenada, which then suffered under the hard fist of the right-wing successor and growing inequality.

33   Panama, 1989

Manuel Noriega, the "Scarface" of Central America, seized power with the help of the United States after a populist leader was killed in a mysterious plane crash in 1981. Within two years, Noriega—a longtime recipient of CIA largesse and a known drug kingpin—was in charge. Reagan's drug czar praised Noriega for his "vigorous anti-drug trafficking policy." Ed Meese, Reagan's attorney general and confidant, interceded to prevent career Justice Department lawyers from investigating Noriega's crimes. These were highlights of a decades-long embrace of Noriega as a useful U.S. henchman. Of course, the entire history of Panama was one of American colonial domination, epitomized by Teddy Roosevelt's peeling it away from Colombia in 1903 to build the Panama Canal, which the U.S. controlled for decades and to which it still retains certain rights.

As to Noriega, somehow along the way he became a bit too ornery even for our own official drug merchants, what with his cocaine business and all, and became an embarrassment and liability. He had to be taken out, and taken out he was, with a spectacular display of U.S. firepower that left hundreds or more civilians dead and more wounded in a wholly unnecessary exercise of U.S. gunslinging. The first George Bush declared four reasons for the intervention—protection of American lives, securing the canal, defending democracy, and bringing Noriega to trial for drug trafficking—all of them transparently misleading or false.

That Noriega was a hideous human being there is little doubt, but he was like so many of "ours" and his utility had been cited time and again by the Reagan regime. It has taken many years for Panama to recover from the trauma of that exhibition of American muscle flexing.

34   Iraq, 1991

America's direct interventions in the broad region we call the Middle East have been few compared with the many in Latin America and the Caribbean, and the two enormous wars in Asia fought in Korea and Vietnam. Two Lebanese landings were the closest we've been to a shooting role in the Israeli-Palestinian conflict, though of course the stout support of Israel is nonpareil. Immense amounts of military hardware have gone to Iran (under the shah), Saudi Arabia, Turkey, Egypt, and others. There was the raid on Libya in 1983. Covert action, of course, has been frequent and nasty and widespread. Oil was at the root of most of it, competition with the Soviets for the strategic advantage of the oil, and ethnic support for Israel. But in terms of actual, boots-on-the-ground shoot-'em-up ventures, we haven't been there much. Until Desert Storm.

The "Storm" did not go so badly. It's what came before and after that gives it meaning. The slavish devotion to Saddam Hussein's survival, only to go sour, and the sellout of anyone who got in the way of that relationship were the elements of the prelude. Then, after Saddam occupied Kuwait, a truly megalomaniacal thing to do, we had the long run up to war, the line in the sand, the new world order, all of that. And finally the great defeat of Saddam in Kuwait and the vanquishing of the Vietnam syndrome once and for all. Yes, there was the "turkey shoot" of retreating Iraqi soldiers, thousands slaughtered. There was the lack of postwar planning that led to the massacre of Shias and the Kurdish refugee crisis. Most of all, there was the isolation of Iraq, the sanctions that led to 500,000 more innocent victims. All of that in one way or another led to the second U.S. intervention in 2003.

Interventions are not merely the landing of marines and the mission accomplished thumbs up nonsense. It is a short piece of history that has roots years before, frequently with the kind of culpability Americans would sooner forget; and consequences, almost always sad and violent, long after. Desert Storm is presented in sharp relief as a success of the first rank—militarily clean, clearly victorious, obviously the right thing to do. But it was steeped in deceit and the most destructive kinds of power politics for years beforehand, and the mortality in Iraq since is a consequence few Americans will think about. The less we think of interventions as quick and isolated from the history that surrounds them, the more likely we will be to think again before undertaking them.

35   Somalia, 1992–93

The Somalia intervention was the exiting president's mission of mercy. Operation Restore Hope was meant to bring some order to the chaos that gripped Somalia, and George Bush, after an active few years at the helm—Panama invasion, Desert Storm, collapse of the USSR—decided to cap it all off with a rescue gig in the Horn of Africa. As many people thought at the time, the choice of timing was peculiar: the crisis had been at a boiling point for more than a year, and 300,000 were dead. The Bush national security team seemed dead set against intervention in Somalia or Bosnia, the other explosive humanitarian emergency. After losing the presidential election to Bill Clinton, Bush reversed course and ordered the intervention. "President Bush and General Powell concluded that liberal humanitarianism would dominate the new Administration," writes an analyst of the escapade. "They also believed that liberal humanitarianists, in control of the White House bully pulpit, would campaign heavily for American intervention in Bosnia. Given the shift in power in Washington and the intensity of mobilized political pressure to respond to humanitarian emergencies, Bush and Powell concluded that if the United States was going to intervene in response to a humanitarian crisis, it would be in Somalia and not Bosnia. Somalia was easier."

It turned out not to be, for lots of reasons. The widely applauded humanitarian impulse there (and later, under Clinton, in Bosnia) came late, not only late for intervention, but for nonintervention. That is, Somalia had been a major recipient of military aid—tens of millions of dollars a year—during the late 1970s and throughout the 1980s. Said Barre, the Somali strongman, massacred thousands of Somalis and generated more of the social dissolution that later came to tragedy, while the United States stood by to gain strategic advantage. Howard Wolpe, an African specialist and former congressman, had been among those urging a different approach in the 1980s. "What you are seeing is a general indifference to a disaster that we played a role in creating," he said. A different approach might have included aid to build a sustainable economy and forgoing structural adjustment policies.

So, in the midst of the crisis we helped create and then ignored, an intervention. It was badly managed and finally went awry under Clinton, who paid mightily for the missteps when some soldiers were killed. What is most poignant about this grisly episode, however, is how it presages the Iraq war begun in 2003. The long, shameless support of despotism for years before. U.S. troops ambivalently received by Somalis. "An overbearing foreign military presence in a country which had been free from colonial rule for only a little more than three decades led to growing resentment," writes Professor Stephen Zunes, "particularly since these elite combat forces were not trained for such humanitarian missions." And the cesspool aftermath, a haven for terrorism and human misery.

36   Big Pharma

It is not always easy to understand the impact of American corporations on the rest of the world. We know about oil companies and the tragic history of the Middle East, but most of us have little grasp of precisely how the major corporations affect life in faraway places. The pharmaceutical industry, an enormous, worldwide enterprise that is among America's largest, is one such example, reaping huge profits with significant aid from the U.S. government, while too rarely serving the world's actual need for lifesaving drugs.

Big Pharma is both a domestic and global problem. It is a classic example of how the American economic system socializes costs—that is, makes society as a whole bear the financial burdens—while it privatizes gains in the form of profits. All corporations engage in this, but the pharmaceuticals gain more than most through the patent system, which grants them exclusive rights to sell drugs they have developed and patented.

There is quite a bit of debate about the economic utility—in addition to the moral implications—of patents. Whatever its merits, the patent system has resulted in colossal profits for Big Pharma, which consistently ranks among the most prosperous industries: in 2004, the top nine American drug companies—those listed on the Fortune 500—made a median profit margin of 16 percent of revenues, compared with 5.2 percent for other Fortune 500 industries. The industry constantly claims it needs the profits for new product development, but in fact it spends far more on advertising and marketing (32 percent of revenues) than on research (15 percent) each year. It is easy to wonder, for instance, how an enormously popular prescription drug like Lipitor, taken to lower cholesterol, can remain so expensive after seven years on the market and 40 million users. Many companies spend a lot of resources developing "me too" drugs, products essentially made already by other drug manufacturers.

Meanwhile, unmet needs go begging. The vaccine crisis illustrates this: the manufacture of flu or anthrax vaccines fell way below projected need because they were not profitable. It is this habit that is so damning—the drug industry is geared toward what is healthy for them, not for us. And where little money is available for drugs, as with the HIV victims in Africa, the story becomes even more shameful.

Worried that developing countries were not honoring "intellectual property rights"—patents of pharmaceutical medicines particularly—the World Trade Organization (WTO) implemented the Trade-Related Aspects of Intellectual Property Rights (TRIPS) beginning in 1994. The Doha Declaration, WTO's trade policy platform most recently updated in 2005, required signatory countries to grant twenty-year market exclusivity on patents. While it allowed for "compulsory licenses" (agreements that obligate multinational pharmaceutical companies to share drug manufacturing information with developing nations) for poor countries to produce generic drugs, very few countries in the global south, with the exception of India and Brazil, have the capacity to manufacture generic drugs.

According to the Global Forum for Health Research, a Swiss foundation, only 10 percent of R&D is spent on research into the problems and diseases that afflict 90 percent of the world's population. In the past thirty years, just 1 percent of new compounds marketed by pharmaceutical companies have been for developing-world diseases. Of 1,393 new drugs developed between 1975 and 1999, less than 1 per cent were for the treatment of tropical diseases. The HIV/AIDS crisis in Africa underscores this dramatically: 95 percent of the 27 million infected people in Africa have no access to antiretrovirals (ARVs), which are shown to have extended victims' lives in Western countries where drugs are available, because they are too expensive. In Brazil, which overthrew the patent restrictions in the interests of compassion, ARVs are widely available and have blunted the lethality of the disease.

The developing world is also home to plants, microorganisms, and animals that are the source ingredients for new drugs, many already discovered in some manner through local traditions of care. However, that local knowledge cannot be protected from use—some would call it theft—by Big Pharma. Simply put, the price of a patent in America is $20,000, a figure few people in an African or Indonesian village could afford to protect their goods. Nor is there hope of contesting or challenging a patent at the exorbitant cost of $1.5 million in legal fees and such. This practice of American firms patenting traditional remedies is stirring opposition in many places, like India, which have a long history of what we call alternative medicine. But the practice of "biopiracy" is difficult to stop, not least when the corporations buy off local collaborators.

Is there nothing nice one can say about Big Pharma? Aren't they also in the business of saving lives through better medicine? In fact, while the corporations often do pay for expensive clinical trials, they are not the source of medical innovation. That accolade belongs to the scientists in mostly university labs. Even here, however, the insidious influence of the pharmaceutical companies is felt: according to a *New England Journal of Medicine* study in 2005, medical schools permit pharmaceutical companies to influence what gets published from industry-sponsored drug studies. Strict standards in some medical schools lead drug companies to invest their money in "relatively permissive institutions." Seven of every ten medical school administrators felt pressured to compromise on standards to gain Big Pharma's R&D money.

So the United States, long viewed, rightly, as a font of medical miracles, is in danger of losing this image as it sinks into the shadow of Mammon. Given the muscle of the industry in Washington, where the drug industry contributed $17 million to campaigns in 2004, this is all hardly news. From prescription drug benefits in Medicare to blocking generic sales from Canada to the maintenance of patent rights in the battle against AIDs, Big Pharma just keeps winning. Everyone else around the world loses.

37   The Weapons Habit

The hackneyed analogies to heroin are still difficult to avoid. America is a major purveyor of armaments worldwide, from the small pistols and rifles that the National Rifle Association guarantees will be available to anyone who wants them, to the U.S. government giving or selling fighter jets and tanks to a panoply of sometime friends and petty tyrants.

We do it for money, for "jobs," for slight and fleeting strategic advantage. We do it because everyone does it. We do it because the arms purveyors inside America give out favors to congressmen and generals like lollipops to tots. We do it because our political leaders lack the wits to promote international security without the rote reliance on weapons.

The United States is now and has been for many years the largest provider of weapons to the developing world. It accounts now for about 40 percent of the overall total, up some from the late 1990s. Russia is a distant second with 17 percent. More than $20 billion in weapons were transferred to the third world in 2004. Of all the weapons the United States exported between 1997 and 2004, three out of every four dollars' worth go to the global south. (These figures, by the way, come from the U.S. government.) The Middle East is where most of the U.S. exports go.

What was delivered? According to the Congressional Research

Service, supplied to the Middle East in 2001–04: "The United States delivered 401 tanks and self-propelled guns, 36 APCs and armored cars, 31 supersonic combat aircraft, 12 helicopters, 347 surface-to-air missiles, and 122 anti-ship missiles." This does not include small arms and light weapons, supporting technologies like computers, items for intelligence or police, and so on. This is just the big stuff. The small stuff does not really get counted.

While total amounts of exports ebb and flow, depending on contracts and long shipment times, the overall trend has been for the United States to dominate the global market.

I traced the export of Sikorsky Black Hawk helicopters, manufactured in Connecticut, to Turkey in the 1990s. Dozens of the exceptionally capable choppers were used widely in suppressing a Kurdish insurgency. But the Black Hawks, and many other U.S. weapons, were used in massive human rights violations of innocent civilians in the course of that conflict and afterward. Similar tales could be told from other parts of the world. President Nixon supplied the shah of Iran with anything he wanted, which was a lot, and President Reagan supplied Saddam Hussein with quite a bit of militarily useful equipment to fight Iran. These things have a way of ricocheting.

Perhaps more troubling, given the nature of the conflicts typically fought nowadays, are the small arms and light weapons that are exported, legally or not. Sometimes these weapons—automatic rifles like the AK-47 and its many imitators—have been shipped to the likes of the mujahideen in Afghanistan by the CIA. In that case, some three million went there, and, after the war ended, could be traced to civil wars as far away as southern Africa. Sometimes they are exported legally to a relatively benign buyer, but then reshipped into a war zone.

The infamous improvised explosive devices (IEDs) used in Iraq by insurgents have their history in technologies like land mines, and were first notably used against the Soviets in Afghanistan by the mujahideen. Hmm. Of course, the United States is opposed to signing the Land Mines Treaty that would start to take these unnecessary and deadly devices—which kill and maim many innocent people annually around the world—out of circulation. What goes around comes around.

But the big systems are where the big money is, and we won't be denied our fair share of the profit making. "We need to educate the [Bush] administration and the American people on selling products in a global economy," a defense aerospace industry spokesman said in late 2005, though he noted that "we're not likely to see new export controls from this administration." In 2005, military aircraft sales were $50 billion, up 7 percent from 2004, while the missile sector generated about $600 million, a rise of 4 percent, and the space sector totaled $37 billion in sales (a 3.8 percent increase). The export business is always sold as a key to a healthy American economy, as the jobs are well paying and the United States has a net plus on trade in this sector.

The importing countries are not always such winners. Military elites and less-than-democratic political leaders love to buy the glitzy hardware that makes them look powerful, even if they never use it. Such aircraft and other systems are expensive and often useless for economic development. So we have many sales to countries that do not need the F-16s or attack helicopters or whatever they're after or may use them in human rights violations. Rare is the case where an actual defense need can be identified. "Offsets" are usually part of the big deals, where the American companies buy goods from the country purchasing the military technology, often goods that are not wanted and not marketable. It's a screwy system that perpetuates all the bad habits of the economies of all the participants.

It is saddening to see the dependency of the cities in America that became hooked on defense contracts. They and their reps will do anything to sell weapons abroad, no matter what the consequences. That they are not making mass transit technology or energy efficient refrigerators or other actual global needs is even sadder, but that's the life of a junkie.

38   Demise of Public Health

Among the ill effects of neoliberal economics worldwide is the gradual but certain decline in health care of the developing world. The system of hospitals, clinics, immunization, training, labs, mobile medical stations, and the other elements that go into prevention, early detection, and treatment—a system that was always underfunded— has suffered a steady erosion and near collapse in the last twenty years. And it came at just the time the AIDS crisis exploded.

The fundaments are not difficult to understand. The economic reform of the 1980s and 1990s that was the ruling ideology of the United States and the international financial institutions that America largely controls treated health care like an industry that should be privatized, and the governments that were responsible for health care services were at the same time expected to shrink. Privatization caused a two-tier system in which "profitable" health care ventures—urban clinics, for example—won out over rural facilities and services that were costly to maintain, such as a reasonably broad distribution of medical professionals. Fee-for-service schemes, largely pushed on poor countries by donors and lenders, reduced the poor's access to health care. A brain drain from public to private health care, and from developing countries to the developed, has also hollowed out the public health systems in Africa and Latin America particularly.

There have been other changes wrought by economic globalization that have an inimical impact on health systems, such as large-scale labor migration and the disruption of social organization that is an important pillar of health in local communities. "Poor social and economic circumstances affect health throughout life. People further down the social ladder usually run at least twice the risk of serious illness and premature death as those near the top," notes a major empirical study of this. "Good health involves reducing levels of educational failure, reducing insecurity and unemployment and improving housing standards. Societies that enable all citizens to play a full and useful role in the social, economic and cultural life of their society will be healthier than those where people face insecurity, exclusion and deprivation."

But the assault on public health systems, particularly by privatization. has been more obviously damaging.

As with most aspects of economic reform, the picture for health care systems is complex, and the news is not all bad. Health care systems in many countries were sclerotic and extremely inefficient, marked by indifference and corruption. Some shaking up was needed. But the market model, which was applied to developing countries as if one size fit all, has rarely achieved anything close to its promises, and often was a disaster. Overall health indicators stagnated. Access was uneven. Many hospitals or clinics closed. Technology improvements helped to mitigate some of these effects, but only partially.

Structural adjustment policies—the wealthy world's insistence that the poor world follow its market model—depleted the public health systems in Africa, limiting access of patients to health care, driving up the number of patients per nurse or doctor, and creating pressures for many nurses and physicians—half of a nation's total in some cases—to emigrate to the United States and Europe. Because public health systems are unique educational platforms as well as providers of medical care, this important function to promote disease prevention, so critical in the fight against HIV/AIDS, was also compromised by the slashes in expenditures. The benchmark study of human cost of the structural adjustment policies, *Dying for Growth: Global Inequality and the Health of the Poor*, notes that "poverty and SAPs have undermined the viability of rural economies, promoted mass labor migration and urban unemployment, worsened the condition of poor women, and left health systems to founder. As a result of these shifts, vast numbers of people in Africa are at risk for HIV infection."

Even the U.S. National Intelligence Estimate, compiled by U.S. intelligence agencies, makes a similar point: "Some remain concerned that the [World] Bank's emphasis on fiscal balance can sometimes have a negative health and social impact in developing countries," says the 2000 estimate. The 2002 version notes that the five worrisome countries "have overburdened and under funded healthcare systems and limited abilities to provide integrated, nationwide programs to test people, track infections, and deliver treatment and education programs," and then goes on to recite a litany of inadequacies, many of them the consequences of the demands of the international donors and lenders.

Another scholar makes a crucial point about the social fabric of Africa in particular and the pressures imposed by Washington policies: "During the 1980s economic policies that emphasized exports, often at the expense of rural farms and small businesses, upset the equilibrium of African communities and probably contributed to the creation of a social ecology favorable to the spread of HIV by swelling already significant waves of migration south of the Sahara." Labor migration is central to the logic of economic globalization, and it is also now central to the dissolution of traditional authority, economic planning, and transmission of AIDS.

While HIV/AIDS gets most of the attention of diseases afflicting the third world (little attention, really, compared with the daily alarms about "terrorist threats" to America, of course), there are many others that are wholly preventable or treatable, yet millions are suffering for simple lack of access to health care. Says one major report: "On average, those living in the world's poorest countries will not live to age fifty. In Africa, the leading causes of death still include diseases such as diarrhea, measles, and malaria. Large disparities in health persist both within and between countries. And the health disparities between poor and rich countries are growing."

The gap between rich and poor is growing. Where have we heard that before? Again, it is as if we can learn nothing from actual examples. The health care systems of Europe and Canada work best, provide the best medical care, and are "socialized." Yet we insist that the developing world privatize its health systems as a condition of getting development aid and loans. Ideology trumps common sense again. Who the winners are in this scheme is never very visible, but the losers—by the billions—are apparent for all to see.

39   Covert Action

As any number of commentators tirelessly remind us, it is a danger-ous world. Some countries, dictators, substate groups, and militia, among other possibilities, set out to hurt other people needlessly. Some defense against these bad actors is necessary, of course, and along with that is a functional intelligence service that can provide some inside information and reasoned analysis.

The Central Intelligence Agency, created from the remains of the American intelligence operation of the Second World War, is the U.S. government's main player in gathering and analyzing information. Along with the Defense Intelligence Agency and other military-based organizations (why there is more than one is a good question), and the National Security Agency, which inter-cepts electronic communications, the CIA is our eyes and ears on the bad actors. It's an enormous set of agencies, with budgets in the tens of billions of dollars annually and vast power to snoop.

Surveillance is a practice with endless temptations. If you're watching a bad actor—Saddam Hussein's regime, for example—you might want to watch the corporations doing business with him, and the corporations doing business with them, until you find yourself watching lots and lots of people who probably don't deserve to be monitored. Or someone higher up makes a comment about a possible act of political violence at a protest rally against, say, globalization, and so you think perhaps the agency should infiltrate that group, or even make a little mischief to discredit it. The history of intelligence operations—all of them, not just American—is a record of precisely this kind of activity. And because it is secret and because it is permanent, civilian control is doubtful and imperfect in any case.

The record of success is not too good either. The end of the Cold War, and the weakness of the Soviet Union, seemed to be a complete surprise to the intelligence services, who generally regarded Mikhail Gorbachev as just another Soviet strongman. The tottering regime in Iran in the mid to late 1970s was visible to all, but the strength of the Muslim clergy as a political alternative was opaque to our spies and analysts. When Secretary of State Colin Powell went to the U.N. Security Council in February 2003 to present our evidence of Saddam's WMDs, it was not only the lying that was staggering, but the lack of information—this is all we had after watching this guy for twelve years?

The CIA is often manipulated by political leaders, the president especially, as was clear during the Cold War, when Soviet muscle was consistently inflated, and again in the Iraq fiasco in recent years. The political appointees know who butters their bread. This manipulation, or incompetence, has cost Americans countless trillions in wasted military spending and misbegotten military actions. It cannot even be estimated how many innocent people have been killed, impoverished, or displaced by CIA bumbling, slavish loyalty, or perfidy.

So the analytical capability, given their resources, is not impressive. But that's not where the shame of the CIA ends. *Covert action* is the euphemism for coups, murder, arson, bombings, incitement to violence, arms and drug trafficking, secret wars, and torture, among other misdeeds. We get a glimpse of it only now and then, and that look is not a pretty sight: Not merely immoral behavior, but usually fruitless and counterproductive. Guatemala, Iran, Indonesia, Chile, Congo, Angola, Pakistan, Afghanistan, Brazil, Bay of Pigs, Nicaragua, Vietnam, Panama, Iraq: you name the foreign policy disaster, and the CIA was there. Training the torturers. Planning the coup. Providing a list for assassins. Setting up the drug smuggling. Choosing the bombing targets. It's a full service agency.

Why is this history of failure and depravity not a cause for a complete revision of intelligence agencies? The national security culture, which includes a very formidable communications capability, simply won't allow it—and that should be telling in itself. This absence of thoroughgoing reform is potentially as damaging to America's effectiveness and standing in the world as the misdeeds themselves. The widespread perception in the world is that the CIA is a rogue actor, behind many evil or merely unfortunate events, and this view is fed by the unapologetic attitudes in the White House and Foggy Bottom and the Pentagon and Congress when some CIA-tinged catastrophe is exposed. Torture? We're protecting all civilized people. Unlimited eavesdropping? Ditto. Running secret wars? All in the name of freedom. Ordinary people can smell the mendacity in these excuses right away, and America is the poorer, and no more secure, for this deceit.

40   Billary

When Bill Clinton was elected president in 1992, there was a collective sigh of relief from many liberals, progressives, and moderates. The bad old days of stagflation, Cold War triumphalism, Reagan's failed revolution, and the rest of the right-wing agenda was finished. (Little did we know.) But what was in store—who was Bill Clinton, what did he stand for?—was still largely a mystery.

It still is a puzzle, actually, as Clinton Homme rolls further into former presidenthood, and Clinton Femme rises to bid for a new residency for the family in the White House. But whatever happens to her, we can say this about their performance over the last fifteen years or so: disappointing. Possibly worse. And much of the world is paying the price.

Now, compared with Bush (either one) or Reagan, I would take Billary any old day. No contest. But for two people so obviously bright, apparently progressive in many ways, and—at least in his case—masterful in the arts of politics, the first Clinton presidency was short on achievements and long on equivocation, lack of courage, and ideational confusion. The next Clinton presidency, if there is one, is likely to be the same.

Consider a partial list of the foreign policy failures or near failures, slow-to-react paralysis, corporate favoritism, minor league appointments, and giveaways to the Republican Congress and the Republican Pentagon: the confusion in Somalia, the tardiness to act in Bosnia, the failure to stop genocide in Rwanda; the free trade agreements that shortchange labor and environmental standards and undercut global equity; the favoritism that fatally undermined the peace process in the Middle East; the inability to stand up to the Likudnik lobby in the United States on Iran; the devastating sanctions against Iraq; the spineless burial of the land mines treaty, the international criminal court, and the Kyoto treaty, the slowness to ramp down nuclear weapons stockpiles. There's more.

The principal failure, however, was passing on a unique moment in American and world history to reshape the world after the end of the Cold War. American power and prestige stood at a peak it had experienced only once before, at the end of the Second

World War. Then, the "wise men" of the Roosevelt and Truman presidencies acted on the opportunity with vision and alacrity, and the results were the United Nations, World Bank, Marshall Plan, etc. One can quibble with these institutions and the intentions of the founders, but the verve and the embrace of opportunity were unmistakable.

Nothing remotely like that emerged from the Clinton White House. He allowed the old battle lines to be redrawn by the supposed clash of civilizations, failed to reenergize foreign aid or development policy, promoted military alliances, caved in to the dissing of the United Nations, and generally handled foreign affairs as a series of discrete problems to be addressed incrementally if at all.

The reasons for this obtuseness are not hard to imagine. The Democratic Party, afraid of its own mildly anti–Vietnam War past, gravitated steadily after Jimmy Carter's similar set of disappointments toward the center, or even the phalangist right, never considering that there were other options that would appeal to the American public (not to mention much of the rest of the world). So Clinton never confronted the Pentagon on its poor response to the "new wars" and its inability to rethink its missions. The relationship with the powerfully emerging Europe was to follow their lead on integration, and to insist on NATO expansion—another "New Democrat" gambit with no reason or reward. Saving Boris Yeltsin was more important than reducing once and for all the numbers of nuclear weapons Russia and the United States had (and still have) in their possession. Relations with Latin America were predicated on free trade and little else. Africa was forgotten, Asia an afterthought but for the financial crisis that luckily did not cascade into a worldwide depression, and the Korea crisis that almost cascaded into a shooting war.

It is difficult to recall a single phrase, a single initiative, a moment of inspiration in global affairs that was of Clinton's making. True, he was hemmed in by a Republican Congress after 1995 (the result of Hillary's failed health care plan), but Reagan had a Democratic House for eight years and rarely lost the initiative. Clinton's own peccadilloes did not help, either, but they were his own after all.

Eight years were squandered, and while that can be said of many presidents, few of them had the skills, the intelligence, and the historic moment for change that Bill Clinton had. His lack of clear principles and dull dependency on decrepit Cold War thinking drained him of this main chance. One could say, in contrast to the disastrous Reagan Doctrine or George W. Bush's war in Iraq, that Clinton at least did not lead us into a catastrophe. But one must have boldness and ideas to make big mistakes; instead, others' catastrophes—Rwanda, Bosnia, HIV/AIDS, mujahideen, global climate change—found him, and he could not respond boldly or effectively.

There is every reason to believe that Senator Clinton would repeat this performance: devoid of firm, identifiable principles, one must govern instrumentally, without a larger guiding vision, and such presidencies can never succeed.

41   Defense Contractors

We used to shoot a man who acted like a dog, but honor was real there. . . . But here? This is the land of the great big dogs, you don't love a man here, you eat him. That's the principle; the only one we live by—it just happened to kill a few people this time, that's all. The world's that way. . . .

Those lines, from Arthur Miller's *All My Sons*, encapsulate the play's moral tale and the morals of the industry around which he built the drama, a military contractor who shipped out defective parts during the Second World War. It is a poignant symbol for the industry, one in which profit has always trumped moral concerns, including, ultimately, the well-being of Americans and foreigners alike.

The defense industry in the United States is a private, profit-making set of corporations, and of course they act like it. They are quick to lay off workers in fallow times—not many of those since 1941—and maximize their profitability at all times. That includes the development of hardware that is not really needed, the sale of weapons abroad, and the endless cycles of upgrades and maintenance and replacements of the same unnecessary technologies.

The aerospace industry is the core of the business, and they have sold a range of hardware that boggles the mind. These technologies are redundant and unnecessary in an age when the main threats to U.S. security come from people with roadside bombs and suicide vests.

And the scale of spending, of course, is stunning. "The $417 billion defense spending bill that President Bush signed in July [2004] is 12 percent higher than the average budget during the Cold War, according to the Center for Strategic and Budgetary Assessments (CSBA), which tracks federal defense spending," reported a Capitol Hill newspaper. "The Pentagon expects to spend nearly $490 billion by fiscal year 2009 as new ships, planes and next-generation fighting tools come off the production lines. That's 23 percent higher than the average Soviet-inspired U.S. defense budget." And that does not include tens of billions more in other agencies that are military-related, including the CIA, the nuclear weapons complex, homeland security, or the full costs of the Iraq war. Not surprisingly, the same report notes: "Shares on Standard & Poor's aerospace and defense index were priced at $255.18 on August 24, up 25 percent from a year earlier and near a fifty-two-week high of $257."

The procurement budget, which is where the new spending is for weapons systems, rose almost 50 percent from 2001 to 2005.

Meanwhile, Congress was looking for cuts in school lunch programs and heating assistance for the poor to pay for these boosts in military contractors and for tax cuts for the wealthy. "In 2001," says another industry report, "when the U.S. Defense Department announced the largest contract in military history, the Joint Strike Fighter project was expected to have an approximate budget of $200 billion. Less than five years later, costs have increased to an estimated $256 billion."

That this is bad for the American economy and equity has long been known. Defense procurement is an expensive way to produce jobs, because the technologies are capital intensive. It does not produce consumable goods, and is therefore inflationary. And of course it places a strain on other spending priorities.

But why is it bad for the rest of the world? An argument persists that because the United States spends so much on its military (more than the next eight highest-spending countries combined), the rest of the world need not, and that they are free riders on the backs of the stability the U.S. military provides. In other words, the "good hegemon," the United States, produces a global public good through this enormous expenditure on its military contractors. It is the kind of argument that cannot be proved or disproved, because there is no "other world" where American power is absent. Two regions where the United States expended tremendous military force—Southeast Asia and the Middle East—are not good examples on behalf of this thesis, however. East Asia is slowly but surely swinging into the sphere of influence of China.

And a major study by the Human Security Center in Canada offers a different and more compelling conclusion, noting "a dramatic, but largely unknown, decline in the number of wars, genocides and human rights abuse over the past decade . . . [and] that the single most compelling explanation for these changes is found in the unprecedented upsurge of international activism, spearheaded by the United Nations, which took place in the wake of the Cold War."

Military contractors have another effect, and that is the need to be global in scope as a way not only of generating profits but sustaining their business to deal with ups and downs in the U.S. military procurement budget. As a result, U.S. relations with many countries are shaped by their ability to buy this expensive military hardware, which they don't need any more than we do. Weapons for actual killing will always be found, so the argument that the U.S. defense industry—particularly the large items like fighter jets—is mainly responsible for fueling conflicts is not true. No, it's more the sheer waste and signs of misplaced priorities that send the wrong signals around the world. What does America care about? The profitability of Lockheed Martin and Halliburton and all the rest.

Even the sense of invulnerability of the "world's only super-power"—just look at the can-do graphics and slogans on the Web sites of the contractors—may have led us into Iraq, and may lead to more misbegotten wars and all the tragedy and anti-American-ism that results.

But defense contractors, who have a tight grip on members of Congress (and employ countless former government officials) and the presidency itself, will not permit the change of course that could correct or limit the damage of these practices. They speak of the "new normal" after 9/11 (of course that means a blank check for them), where the supposed complacency of pre-9/11 thinking is no longer permissible. Never mind that very little of what the big contractors sell has anything to do with defending against terrorism, or winning the war in Iraq, or changing the world in ways that would lessen the possibility of security threats to the United States.

In fact, one can plainly see the "war on terrorism" as a way to allow the defense industry to burrow into the U.S. Treasury more than ever. The absence of any debate about a military budget that is larger than the peaks of the Cold War is a sign of how completely the industry has triumphed, in part because, in many respects, it *is* the government. And that is the most troubling fact of all.

42   "We Don't Do Body Counts"

When U.S. General Tommy Franks uttered those words in 2003, he was conveying the new sentiments of the American military and its civilian leadership, that counting the dead of "the enemy" was not necessary or useful. Franks, who may be remembered as the only general in the annals of American history to lose two wars, was simply repeating what his political handlers told him to say, as all active duty generals do. In this case, it was an attempt to deflect the moral consequences of a "war of choice," a lesson Franks's generation learned from Vietnam. But the "no body counts" policy reverberates around the Arab and Muslim world, to America's detriment.

The policy is an insult and a mistake for two reasons. First, it lends the impression—or is it a fact?—that the United States does not care about civilian casualties. In the autumn of 2005, in a fairly typical sequence, the military announced that a sweep of Anbar province in Iraq had resulted in the death of 120 "terrorists." No civilian casualties were reported by the U.S. government, or by the American press. Al Jazeera, the Arabic news organization, had firsthand accounts of dozens of casualties. And it is inconceivable that major military operations of that kind would not result in casualties of the innocent. This is an embittering legacy of the war: not merely the fact of large numbers of war dead, but the neglect of even acknowledging that this could be occurring or is important enough to investigate.

Second, it is bad for the war effort itself. The American people have a right to know what is going on in their name. Learning about things like Abu Ghraib and casualties from foreign news sources or NGOs makes the revelations all the more troubling, as they think they are being lied to by their government. (Which they are, of course.) And military planners themselves should understand what the effect of operations is on civilian populations. Family ties are strong in Iraq, with close extended kinship networks; killing of family members, especially innocent family members, is likely to produce more resistance—and more terrorists. It is one of the seemingly inexplicable things in Iraq—how could the insurgency grow when America is so clearly a liberator, where even Sunni Arabs will ultimately be better off if only they would lay down their weapons? The answer is not only that they are former Saddamites or jihadists. The far more probable answer is that the insurgents are driven in part by acts of defense, in effect, or vengeful honor.

A military officer told me around that same time that "rules of engagement" for U.S. troops were so broad that civilians even faintly suspected of being insurgents were routinely "blown away." Men talking on cell phones, for example, while a U.S. military convoy was passing were fair game for shooting. Many anecdotes of this kind circulate, but have stimulated little curiosity on the part of journalists.

Most take at face value the estimates of Iraq Body Count, a noble effort to count, via press reports, the total number of Iraqi civilians killed in the war. Their estimate by the end of 2005 was about thirty thousand, but their method was incomplete, as they readily acknowledge, since they count only those who are reported dead in two or more reputable news sources. That's like doing the census of the United States by counting everyone mentioned in the news media.

A more complete estimate was provided by a team of epidemiologists, led by American and Iraqi health professionals, and published in the British medical journal *The Lancet*. Using a well-tested method of random cluster surveys, interviewing more than 7,000 people, their midrange estimate was 98,000 dead in the first eighteen months of the war, with 80 percent of those likely to have been killed by U.S. and U.K. forces.

That report was widely dismissed in the United States as politically motivated or flawed, though the secretary of state and many others used the same method to estimate casualties in other wars, such as the Congo. (The method, by the way, while widely misunderstood, is perfectly sound.) The violence, by most accounts, increased in the next eighteen months, and one can safely assume that the actual dead in Iraq now exceed 100,000 by perhaps tens of thousands more.

The real reason why *The Lancet* study is ignored, and the whole topic of civilian deaths downplayed, is that that scale of mayhem is just too sickening to accept in a news media that largely supported the invasion, and by politicians who would pay a price for even indirectly criticizing the conduct of U.S. troops who, after all, do the killing.

The moral consequences of war are always inconvenient. They are especially troubling when a war has a veneer of righteousness. This attitude afflicts the media elite as much as the political leadership. "We don't do body counts" could have been uttered by the editor of the *Washington Post* as easily as the general in charge. That they are both morally bankrupt on this issue is obvious for all the world to see.

43   Getting High

Much of my professional time is spent studying armed conflicts around the world.

One can't help noticing that wars today are often mixed up with crime, and that crime is often about drugs—heroin and cocaine, in particular. The production of opiates is connected to the wars and instability in Afghanistan, Pakistan, and Burma. Cocaine is produced in the Andean countries of South America, particularly Colombia, Peru, and Bolivia, and all three have suffered from ongoing civil wars—Colombia's is almost forty years old—and social unrest. Then there are the transit countries, like most of Central America. You add it up and a lot of countries are involved. Of course, one country is most involved, not as an exporter, but the consumer: the United States.

Getting high is an American tradition. Alcohol and tobacco consumption is as old as the Republic. The use of legal pharmaceuticals for depression, anxiety, sleeplessness, and the like increased markedly after the Second World War. The legal drug market set the stage for illegal drug consumption. Even now, after all the public service campaigns about these issues, the consumption of alcohol by American teens and preteens is astounding in scale: 20 percent of the alcohol imbibed nationwide goes down the gullets of kids between twelve and twenty years old, and in that age group, half of them drink. They account for $22 billion in booze.

So the stage is set for illicit drug consumption. Americans consume more cocaine than any other country, 300 million metric tons annually. In the 1990s, about $70 billion was spent in America on coke by 3 to 4 million "hard-core" users and some 6 million occasional users. Up to a million Americans were hooked on heroin, and that cost about $20 billion a year. Trends suggest that use of cocaine may actually be declining, but statistics in general are a little dodgy when it comes to these practices. It's still a very big business.

The industry that supplies the habits of Americans gives new meaning to the word *globalization*. West African couriers go to Bangkok to purchase Burmese-made heroin and run it through no-hassle airports in Africa and take their chances at border crossings in Mexico and Canada. Cocaine shippers, we know, have their own air fleet. The transit points for all this stuff read like a who's who of failed states (or venues of the Reagan Doctrine): Angola, Cambodia, Guatemala, Nigeria, Honduras, Mozambique.

The drug money—a little goes a long way in some of these countries—feeds the corrupt and brutal, rogue cops and dirty politicians, ready to take the graft in one hand and U.S. "war on drugs" money in the other. They are often involved with the other contraband that fuels war and crime: gunrunning, diamonds, or even higher-end goods like nuclear technology. They sometimes have connections to the likes of al Qaeda. It all seems to go hand in hand. And drugs are at the center of it.

The war on drugs is generally considered to be a failed policy, and an expensive one, though it has its defenders. Our federal and state governments spend something like $50 billion annually, both at home and abroad, in the drug war. A million are arrested, many of them incarcerated, bringing on more billions in costs. In places like Colombia, the war on drugs is mixed up with the civil war itself. Local police and military elites use the drug war for other purposes—not only old-fashioned graft, but as a way to settle scores and dispatch enemies. Eradication of crops only works if the local people want it and there are alternatives, which are rarely in the mix. Very few independent analysts regard the war on drugs as a success, mainly because it is being fought in the wrong places—the problem is not abroad, but in ourselves.

Free trade helps the drug trade. The war on terrorism may hinder it in some places, but help it in others. In Afghanistan, the overthrow of the Taliban opened the door to new cultivators and exporters of heroin.

It's a very confusing picture. The one remedy that has not been tried, of course, is legalization. There is a ferocious debate about the harmful effects of drugs, and what legalization (controlled, taxed, etc.) might bring. But one thing is certain: the hunger for illegal drugs in the United States reverberates around the world. It is violent, corrupting, and enormously costly to millions of people on every continent.

44 Torture

I will keep this one short, because it is so obvious and hardly any rational and moral individual would disagree with me. In fact, there is so much unanimity on this matter among knowledgeable people worldwide that I thought perhaps this should *not* be one of the *100 Ways*. But then I saw Condi Rice in Europe defending the "renditions" of "suspects," spirited off to secret prisons where no doubt they would come in for some serious hands-on interrogations, claiming these contemptible practices saved European lives—almost certainly a complete falsehood—and I thought, well, yes, torture deserves a few words.

America has overall been quite free of torture as an official state policy or practice, so it is perhaps a little premature to claim that the recent reliance on torture prisons for the massive detentions of fighters from Afghanistan and Iraq and others has "screwed up the world." Too soon, but not too far fetched. The revelations about the U.S. torture techniques and the persistence of the Bush administration in defending and using them are a colossal national shame that has muddied whatever conceivable moral clarity guided the new crusades in the Middle East.

Apart from being morally repugnant, a slap at the ideals the country tends to uphold, and a violation of international law—often flimsy reasons in the minds of torturers—the practice of maximizing pain doesn't work. People who actually know something valuable (unlike the thousands of low-level prisoners at Guantánamo and other prisons) are the least likely to talk. And some will talk and say anything to stop the beatings, burnings, poisonings, and other methods in the torturer's quiver. Hence, the many false alarms and "orange alerts" since 9/11 (which, conveniently, also have political value). "No one has yet offered any validated evidence that torture produces reliable intelligence," notes General David Irvine, a specialist on interrogations. "While torture apologists frequently make the claim that torture saves lives, that assertion is directly contradicted by many Army, FBI, and CIA professionals who have actually interrogated al Qaeda captives."

In its Eight Lessons of Torture, the Center for the Victims of Torture, an experienced, Minneapolis-based NGO, notes in lesson number one (that torture does not yield reliable information) that "nearly every client at the Center for Victims of Torture, when subjected to torture, confessed to a crime they did not commit, gave up extraneous information, or supplied names of innocent friends or colleagues to their torturers." And as many people have argued, including former interrogators, torture has a corrupting influence on the torturers.

The big "what-if" in the debate is "what if a captured terrorist knew of a plan to detonate a nuclear weapon in Manhattan, should we use all means to stop that?" Such what-ifs depend on many implausible scenarios converging. The simple fact is that suicide bombing has shown that the most politically violent people will die for their cause; this is not exactly news. So if in the unlikely case (getting a nuclear device is an extremely low-probability event) someone did know of such a thing about to happen, and we're pulling out their fingernails, we can rest assured they won't talk, because they would has committed to dying anyway. People who argue otherwise are not only morally corrupt but naive. We could, however, pose a more likely what-if: What if a would-be terrorist becomes a deadly fighter because America is torturing his compatriots? That what-if is already under way.

Some people—American torturers—have blood on their hands as a result.

Case closed. The torture, the illegal detentions, the unnecessary killings, the grisly prisons—not a single benefit has been shown from this tawdriness and moral depravity. It is likely to outlive its alleged purposes and brand the perpetrators forever.

45   Consumerism

We're all committed consumers now. It's almost impossible to avoid. From the three-dollar cup of coffee to the new car every three years to the supersized house, we can't get enough. It's all about getting and having.

I used to wonder why shopping was so popular a pastime. There is the satisfaction of buying something, this I know. But shopping habits far exceed the time needed merely to purchase items, even things we don't need. No, shopping is a form of fantasy life. How would I look in that suit? Would that sofa be fun to have in our family room? Aren't those dinner plates elegant, perfect for the elegant dinner parties that would surely result from purchasing them?

And so on. Going shopping in a good mall or shopping district is to engage in a series of these daydreams. The malls and districts cater to this, of course. There is little to distract you from the fantasy, apart from fast food courts, which are instrumental to keeping us in the mall. It's often said that shopping malls are the new town square, but in a real city, there are many things in addition to shops—homes, schools, community centers, government and commercial offices, food stores, entertainment venues, parks, and quirky or sheerly utilitarian stores, among other things. Shopping malls allow only stores, really, and a narrow band of those.

The more upscale malls—Camelback in Scottsdale, Georgetown in Washington, Copley Place in Boston, etc.—allow few hardware stores or pharmacies, which offer not many fantasies.

This is not exactly new, of course. "Seek among men, from beggar to millionaire, one who is contented with his lot, and you will not find one such in a thousand," wrote Leo Tolstoy. "Today we must buy an overcoat and galoshes, tomorrow, a watch and a chain; the next day we must install ourselves in an apartment with a sofa and a bronze lamp; then we must have carpets and velvet gowns; then a house, horses and carriages, paintings and decorations. . . ." Aristotle believed that the "avarice of mankind is insatiable." Thorstein Veblen, the brilliant sociologist of the early twentieth century, wrote of the "conspicuous consumption" of the wealthy, and how such consumption established social hierarchy—as true today as then, visible in the Hamptons or Beverly Hills or Pacific Heights.

But what is interesting, and a little disturbing, is not how conspicuous consumption continues to afflict the silly rich, but how it now infects a much broader swath of American society and indeed the rest of the world, that part of the world, at least, where consumerism is conceivable. And what one sees in America, and increasingly elsewhere, is an ethos of consumption that really did not exist a couple of generations ago, when the norms of the middle class dictated thrift and austerity, living within one's means, saving for the future and for one's children, and not being conspicuous, really, about anything. There was a stultifying conformity in that, perhaps. But the thrust of it was not all bad.

American consumer culture actually took off when the goodies long reserved for the well-to-do were made available in bastardized form for everyone else. Once, good restaurants were not something ordinary middle class or poorer people indulged in; today, going out to eat is commonplace, and restaurant culture, from Starbucks to the Cheesecake Factory and up, is available to everyone, everywhere. Cheap imports, including cars, clothes, electronics, and so on, have made it possible for people on the lower end of the economic spectrum to buy things that were unimaginable a few years before.

Fueling this purchasing are easy credit, ATMs, and television advertising and the popular culture that both extols and makes fun of wealth. Much of buying is competitive consuming, as the sociologist Juliet Schorr observes, among people of roughly similar incomes and backgrounds who aspire to a step up the social and economic lifestyle ladder. The reference group is typically those in the upper 20 percent, who set the standard for consumption, and that standard is blazing away at us from TV sets, magazines, and all manner of advertising. The workplace itself became a venue for learning about consumption patterns and desires.

Keeping up with the Joneses, much less aspiring to be no better than your neighbor, gave way to much bigger appetites. Affordable luxury. Living well is the best revenge. You can't take it with you. By the mid-1990s, the middle classes were spending much more of their income on homes (and second homes), and house size had doubled from fifty years earlier, for example, and spending on premium foods and entertainment had skyrocketed. Credit card debt had grown too, while savings rates dropped precipitously. People were going into debt to buy large TVs, large SUVs, large houses in sprawling suburbs. The average American was discounting the future in order to consume today. Nearly half of American families spend more in an average year than they earn. By the mid-1990s, Schorr concluded, "many households felt pessimistic, deprived, or stuck, apparently more concerned with what they could not afford than with what they already had."

It can be argued that consumption is normal desire and that a successful economy has made it possible. Why the bitching? We don't need all this stuff, not even close. Consumer desires are fabricated, not natural. The only reason to relentlessly stimulate these consumption habits is to make a buck, not to make people better or happier or safer. Ordinary folks are going into debt and leaving little for their communities or children. (Tax cut fever is supported by the middle class mainly to increase consumption.) The mountains of waste increase. The imports of cheap stuff are hurting our long-term economic stability as a country and not doing enough for third world development. It's circular bad behavior, seemingly innocuous, but in the end enormously harmful.

And what's popular in America almost always catches on in other places. Europe is consuming in similar ways nowadays. The copycat countries are delving into the same nonsense, with the same results. But even on our own, the consumerism that has been cultivated in America for decades is harmful to the rest of the world. Consumption patterns of energy resources are most obviously harmful, but even arable land and other natural assets are wantonly drawn down, and someone has to pay. Eventually, everyone will, and the price will be high.

46   The Attack on Science

Of America's many assets, its scientific acumen is among the most golden. Along with bountiful natural resources and a skilled, diligent workforce, the United States features an astounding array of the highest quality research universities, private labs, and government-sponsored institutions are the best the world has ever seen. If you've had the privilege of working in major research organizations, as I have, you know the simple fact of America's vast abilities in virtually every field of scientific endeavor.

This science has helped save lives through improvements in medicine and nutrition, fueled economic growth, and brought a number of other wonders into our lives. Scientific institutions do not have an unblemished record, of course. Big science, especially, that is geared mainly to produce profits or instruments of destruction, has contributed its share to the world's ills, some of which are documented in these pages. But science as a pure concept, namely, the pursuit of truth and its application to solve human problems, has added immeasurably to human existence, and America has been among the leaders in all fields.

One would think that the right wing in this country would be science's biggest booster. The connection to economic growth and technological development is quite direct. Scientists by and large are politically neutral. Technical mastery is the fundament of U.S. military strength. And the political left has long been a critic of science, everything from nuclear energy to genetically modified foods.

But the right wing must have its cake and eat it too. The Bush administration and its notorious "base" have mounted an assault on science that is truly breathtaking in its scope, nuttiness, and consequences. And because the United States is not only a world leader in science, but the eight-hundred-pound gorilla in the world's living room, the right-wing attacks on empiricism, rationalism, and the Enlightenment are taking a terrible toll.

Where to begin? The nuttiness, which on the surface seems least harmful, is easy to dismiss as the work of religious cranks. The mind reels when reading about their relentless mugging of Darwin and the theory of evolution. (The mind reels even more when seeing opinion polls showing that even many liberals say it's okay to teach "intelligent design" in schools, signaling a failure of scientists to dispute the anti-Darwin nonsense.) While it is likely that this is a passing fad of the Christian right and the exorbitant attention they are given in the news media—aided by a president who endorses their position, of course—it is nonetheless a tragedy for the thousands or millions of schoolchildren who are being taught phony ideas dressed up as a scientific debate. So while it would be comforting to dismiss creationism and its progeny as the work of religious nutcakes, those nutcakes occupy many executive mansions and statehouses.

More important for the world today (since most of the world sees creationism for the laughable lunacy it is) are assaults on science in two very serious forms. The first is the growth of "know-nothing" dissembling. The second is the more pernicious goals of Christian fundamentalists, as with their victory in retarding stem cell research.

In early 2004, the Union of Concerned Scientists, a leading public interest group, detailed many of these abuses in a report that has been endorsed by many of the nation's leading scientists. Among the report's findings were numerous attempts by the Bush administration to slant or distort scientific studies of environmental issues, such as changing the contents of a government report on lethal mercury pollution from coal-burning plants. The most serious of all the abuses are those surrounding climate change, in which George W. Bush and the usual suspects in his circle deny that the science proves that human-caused changes in climate warrant serious policy changes. All this mendacity is in the service of the oil industry and the other slaves to the short-term bottom line, as everyone knows. But the lack of attention to this issue in the news media (there are nearly as many mentions of Paris Hilton as of climate change in major newspapers) makes it difficult for average citizens to sort out the false claims from the real science.

I had long thought that the Christian right's opposition to stem cell research was based on its religious dogmas and the need to raise money from naive members of their congregations (not everyone, as it happens, is homophobic, so other causes are needed). As you know, stem cell research is possibly the most promising avenue in medical science today for developing cures for horribly disabling maladies, like multiple sclerosis. Embryonic stem cells are needed to fully exploit these amazing possibilities. The right has successfully blocked federal funding for new stem cell lines of this kind through an executive order that Bush issued in 2001. Other countries are moving forward; our scientists will likely be left behind. Cloning stem cells would be equally promising, but the extremist Christians oppose that too, even though there are no biblical injunctions against it.

It turns out that this religious nonsense also has some basis in corporate greed. While a number of major pharmaceutical firms would benefit from the therapies produced by this new research, they would also lose billions of dollars in revenues from current medical treatments. A good example is that of diabetes. If stem cell research could devise a simple way to coax the pancreas to produce insulin the way it is supposed to, the cumbersome and expensive treatment of diabetes today, with its glucose monitoring, insulin injections, and so on, could be eliminated. Twenty million diabetics in the United States and many millions more worldwide would be freed of this burden, and cost, which now produces a multibillion-dollar industry. Reportedly, some of Big Pharma are backing the religious right's attack on stem cell research.

Because so much of what independent science brings us is bad news for religious crazies and corporate elites, it has increasingly been targeted as a morally corrupt enterprise, or compromised by military secrecy and waste. No wonder other countries are catching up or surpassing us on patents, scientific journal papers, and agenda-setting research. We sustain scarcely half of U.S. patents; in the publication of physics papers, Western Europeans swept by Americans a decade ago.

The great contributions science makes to human comfort and longevity, and the earth's sustainability, are now going to be delivered by others.

47   The Failed Presidency of George W. Bush

Failures are sad things to behold. No one likes to be a failure, and no one wants to be associated with a failure. With countries, failure is regarded with special alarm and opprobrium. One such category of concern—"failed states"—says it all: they are the road-kill of the global community. It's not a pretty sight.

Many of the forty-two occupants of the White House qualify as less than successful. But it counted for little to the rest of the world if James Buchanan or Ulysses Grant or Chester Arthur were failures. It matters a lot if the current president is, and so we take up the particularly onerous chore of declaring and explaining the failed presidency of George W. Bush.

This, I hasten to point out, is not a partisan exercise. The *100 Ways* is rigorously nonpartisan, as you have noticed and no doubt appreciated. We have revealed the deficiencies of the Clinton White House. One could add to that Jimmy Carter, LBJ's Vietnam policy, and other foibles of the Democratic Party. It just so happens that George W. Bush is president today, and his presidency is such a striking example of failure. We need not wholly blame the person of George W. Bush, of course. We can blame his mother, or Antonin Scalia, or someone else if that seems less harsh. But the facts are difficult to hide.

The indictment includes the war in Iraq, which has resulted in massive loss of life and general mayhem and chaos in the region, all perpetrated on the basis of failed intelligence and the failure to recognize how contrived the intelligence was; the failure to find bin Laden and put an end to al Qaeda, instead swelling its ranks with new terrorists as a result of the war in Iraq; the failure to address the coming catastrophes of global climate change; the HIV/AIDS policy driven by religious extremists, costing God knows how many lives; the rapid expansion of the national debt, after having started in 2001 with a gigantic surplus in the federal budget; the equally alarming expansion of the trade deficit, to new records; corporate cronyism run amok; and the attack on civil liberties at home and abroad that even has some conservatives in an uproar. And those are just the highlights.

Why does this matter? Haven't we gotten by with lying, bumbling, incompetent, bankrupting, belligerent, and generally embarrassing chief executives in the past? There is, true enough, a long list, and the harm can be viewed as cumulative. But "W" set out to be a compassionate conservative who has instead acted consistently as a coldhearted destroyer, blundering on with a vast recklessness that has reached every corner of the globe.

Apart from the specific consequences of misbegotten policies, many of which are detailed elsewhere in these pages, the failure can be symbolized by one day: September 11, 2001. Not that Bush failed to stop the atrocities committed that day in New York and Washington (although there's a case for that too), but how "9/11" as a package of policies, maneuvers, and emotional manipulations has been used time and again, without shame, to further Bush's political career and a revanchist agenda at the expense of America's proper role in the world.

This is the most telling *moral* failure among the lengthy bill of particulars. This could have been an occasion to unite nations in common cause against the darkness that is al Qaeda and the septic ideologies it espouses. Instead, it was "if you're not with us, then you're with the terrorists," and Guantánamo and renditions and Abu Ghraib and detentions of American Muslims and wanton killing of civilians in Iraq; a world, once deeply respectful of America, increasingly despises everything we stand for. He could have seized the moment of national unity and mourning to demonstrate compassion at home and honor the aspirations of the people who pay for and fight his wars, but instead has set out an extremist economic juggernaut of giveaways to the rich, to the oil companies and defense contractors. His calls for freedom are shadowed by his insistence on secret government, thousands of arrests and imprisonments without due process, and relentless spying on the American people.

This is failure of a very special kind. It is not mere incompetence, although much incompetence is apparent for the world to see. It is not just bad judgment, though bad judgment is rife through the administration, from the vice president to two secretaries of state to the perplexing spectacle of the secretary of defense and his cohort. And it is not just selfish conservatism, because a true conservative would preserve rather than disrupt, coerce, and socially engineer the rest of the world. It is the failure of a small clique of would-be imperialists, men and a woman who play war games from the safety of their offices and drain the national treasury for cronies and the class to which they seek membership.

It would be amusingly sad if it were not causing such misery around the world, such anti-Americanism, so many bad examples of governance. America will one day look back on the bloody chaos of the Middle East, the mountains of debt, and the lost time to reverse climate change, and ask—what were we thinking when we gave George W. Bush a free pass?

48   Liberal Hawks

For at least three decades, a species of political animal has garishly flapped its wings and left its markings around the world— America's sui generis liberal hawk. Beginning perhaps with Henry Jackson, a pro-labor but fiercely anticommunist Senate leader, the liberal hawk has popped its head up every time an international crisis erupts. It's been the bane of liberalism and has robbed hawkishness of its essential militaristic flavor.

Liberal hawks were common during the Cold War, when being anti-Soviet had no political consequences, but being thoughtful about the insanity of the rivalry did. More recently, the liberal hawk has fluttered noisily about Iran's nuclear ambitions, Putin's tactics, Hugo Chávez's bluster, Islamic terrorism, antiglobalization marchers, and those old standbys, Saddam, Arafat, Assad, Castro, and France.

The idea animating the liberal hawks is that they—perhaps green, feminist, multicultural, small-d democrats—can also be tough when it comes to defending America and its ideals. They excuse no dictators. The banner of political liberty must be raised above all others. The spread of democracy and the smiting of tyrants can and should be the work of the Upper West Side, Cambridge, and Berkeley every bit as much as it is of Wall Street, Southie, and the OC.

Who can sneer at human rights, liberty, and democracy? Not I. But when I see a liberal hawk, I smell a rat. And the stink comes from—how to put this delicately?—both expiation and exhalation. The first draws upon a religious doctrine having to do with guilt: the way to make up for sins is not to inflict punishment on oneself, but on others who consent to this substitution. We are guilty (as liberals) because we are weak and accommodationist, it would seem, but as reconstructed liberals we can punish those leftists who are not. (It is "consent" in that liberal hawks always get the press.) The exhalation part is simpler, constant, and from all ends. But like air that comes from the body, it has been exhausted of oxygen.

And the absence of oxygen is indeed suffocating. The most recent, and perhaps the most notorious, gathering of the flock was to urge the U.S. military to invade and occupy Iraq in 2003. The arguments had to do with the now laughably deceitful WMD hoax, or Saddam's ties to al Qaeda, and the other White House canards swallowed whole. Most of these liberal hawks argued later that they were really interested in human rights in Saddam's Iraq, and liberating twenty-five million Iraqis from his grip. Fair enough, but was a U.S.-led war the way to achieve that?

There has been little from LH perches about the extraordinary level of violence in Iraq, now approaching 200,000 dead. Nor has there been much thoughtful said about the de facto division of the country, the massive crime and corruption, the new training grounds for jihadists, the victory of Iran in the south, and other, similar wrinkles. This is the victory of liberal principles?

The retreat of the liberal hawks has been that reliable foil, which is that Bush somehow messed it up. It *was* a grand enterprise, but the arrogance of the neocons is the real reason things went awry—not, heaven forbid, that it was a bad (illegal, immoral, bloody) idea to begin with. Not that it sets a horrific example and precedent. Not that it is imperialistic by its very nature. Not that many many people were going to die. Not that other important issues would be shrouded by the relentless pursuit of war. No, somehow knocking off Saddam was a way to prove that we, too, can be tough guys, and nothing else mattered—not law, politics, human security, dignity, or that favorite catchphrase of 2002 and early 2003, "moral clarity."

When principles are abandoned, nasty consequences usually follow. Liberal internationalism once had a proud pedigree, sullied and bastardized by the liberal hawks of the Iraq fiasco.

49   The Puritanical Ethic

The Puritans set out for the New World from England. The "Puritan ethic," the salvationary, hardworking spirit of those early

Protestants, may have been coined by German sociologist Max Weber. But the *Puritanical* Ethic, the supposed moral uprightness of no-fun-at-any-costs, is all-American, and it is spreading across the globe.

Puritanicalism has fought a pitched battle to save the souls of the world. It brooks no nonsense when it comes to the things that are generally associated with having fun—drinking alcohol, smoking tobacco or marijuana, enjoying sex, gambling, telling off-color jokes, swearing, and other such pleasures. Some of these things are bad if done in excess, but eating ice cream is bad for you if done in very moderate excess, and I have noticed no attempts to shut down the ice cream industry.

The threats to the nation's morals in these depraved deeds derive from the libertine qualities associated with them. And libertines tend to be nonconformist. Enjoying sex might challenge certain attitudes about lifetime monogamy and the institutions that uphold it; and homosexuals famously enjoy frequent sexual liaisons. Alcohol and marijuana render one silly and unproductive (*and* sexually promiscuous, a trifecta of depravity). The evils of gambling hardly need explaining.

Focus on the Family and other self-appointed guardians of public virtue will scream at the sight of a naked reproductive organ on television, but it's verboten only if that organ is seething with desire. If it's being blown to smithereens by an American fighter jet in a Muslim country far way, that's another matter. The disparity between extolling and celebrating violence and repressing and demonizing sexuality in America scarcely needs repeating. But it's worth noting how much more energy is put into blocking the sight of a breast on television than drawing attention to poverty, environmental destruction, war, or racism.

The righteousness is so unrelenting that one suspects ulterior motives. In Maryland, where legislation to legalize slot machines at racetracks was on the table, a fierce opposition rose to smite it down. No matter that many forms of gambling exist, including the illegal kinds that can get people in trouble. No matter that surrounding states had slots, draining tens of millions of dollars from Maryland. No matter that the bill, by helping the racing industry, would have saved thousands of acres of pasture. No matter that the slots revenue would have largely gone into public education . . . aha! There's the rub. The Christian right loathes public education, with its secularism, evolutionary theory, and who knows what other evils. And, sure enough, the members of the main coalition to stop the slots bill were perfervid Christian puritanicals.

Fortunately, the Puritanical Ethic is spreading only slowly across the world. Europeans, notoriously libertine (didn't they *invent* immorality?), are succumbing to the antismoking fervor that began in New York City. Now Dublin and London ban smoking in restaurants. Banning sexual content from television is gaining adherents, even in Scandinavia. Australia's censoriousness is among the most severe in the industrialized world.

America's contribution to this is sometimes subtle and self-contradictory. Its porn industry is the largest globally and exports its product profitably. The U.S. manufactures and exports cigarettes, and indeed is the world's leader. It is an eager purveyor of alcohol to the world (alas, only the eighth largest). American gambling corporations are building supervenues in Macao and elsewhere. (Coincidentally or not, the leading exporting states for each of these vices are now governed by Republicans.)

At the same time, the U.S. government is promoting sexual abstinence, rather than condom use or low-cost distribution of prescription drugs, to deal with the HIV/AIDS crisis in Africa. So American pornography can fuel fresh erotic desire, American alcohol can loosen inhibitions, American tobacco can enhance the pleasure of both, but sexual abstinence is the answer to all these stimulants. That's the Puritanical Ethicists' special form of logic.

One wonders: Will nostalgia for a time of freedom of choice, unencumbered by rampant hypocrisy, be outlawed as well?

50   Democratization

The Initiative for the Greater Middle East, as President Bush grandly called it, was at heart a project to bring democracy and all its blessings to the Arab lands so long under the yoke of this or that colonial power or authoritarian regime. A noble idea, it tracks perfectly with the crusading spirit of America, and Bush especially, and the only broadly acceptable basis for that kind of "we know what's best for you" attitude is the introduction of democracy. Who could object?

Democracy has been on the march since the Second World War, supported by every president and every Congress, as well as, dare we say, Europeans, the United Nations, and other agencies of social change. This is a good thing. People should be self-governing. But the fly in the ointment here is not everyone's endorsement of democracy, but defining what it is, when it can be introduced (or imposed), how you know when it is working, and whether that is the full purpose of political development. It is "democratization"— the ideology of political intervention—and not democracy itself that is troublesome.

There are two problems with this impulse. The first is the zealousness of the enterprise. The second is its narrowness. Both work against the likelihood of digging deep, nourishing roots for democracy to thrive.

We can understand the evangelical exuberance Americans bring to the idea. We revel in our democratic spirit, institutions, and processes, rightly so for the most part, and seek to share that with the parts of the world that are not so lucky. In the last several years, however, this has led to some rather bizarre proposals, like invading a country and participating in an extremely ugly and unpopular war and occupation. Zealotry can lead to extreme actions that require steadfast adherence, not just quick and easy military invasions. It can and often does embrace the support of guerrillas and the idea that fomenting violence is a good path to democracy. That violence and violent actors, instruments, culture, and the like are ultimately destructive of political comity is not often considered in these ill-formed plans of action.

The democratic fever can also be deafly insistent that our way is superior to other forms of popular sovereignty. It would be instructive, for example, to see the reaction in America if a European said we were not upholding democratic rights because there is no universal health care in the United States. An articulation of rights, after all, comes with the territory, and universal health care or full employment could be legitimate, even compelling, standards of democratic practice. Equity, fairness, and social caring can be—and are—as important as elections in democratic practice; our European, Japanese, and Australian friends, to name a few, have succeeded in this far more than we have.

The last point signals the problem of narrowness. We tend to regard democracy as the right to choose our political leaders at the ballot box. Some might throw in the rule of law and functional political parties as key elements of democracy. A number of policy makers have also included market-based economies, openness to foreign investment, and respect for property rights.

This menu is very much American cuisine. It is geared toward the kinds of democracies we want to supplant the old socialist regimes, whether in the former Soviet Union or the Arab world. We want rule of law to order everything, to make sure contracts are respected and corruption is minimal. We want political parties of a certain kind—that is, committed to the other values on the menu. (Major U.S. conduits for "democratization" funding are the International Republican Institute and the National Democratic Institute, which are—surprise!—run by the Republican and Democratic parties. Not much room for innovation there.)

Most of all, we want market economies, because without that, well, it's not really a democracy, is it? But market-based economies are not very democratic themselves, as they tend to be ruled by and for those with ample resources, be it capital or oil or something like that, and very often those with the resources are from somewhere else . . . like America, for example. One can have market economies without democracy, as China seems to be proving with a vengeance, and democratic political systems with a socialist agenda, as much of Northern Europe has managed to do quite successfully.

Rapid political reform, including democratic processes, can even be destabilizing and lead to civil wars—particularly when marketizing policies are also imposed—in societies not strong enough to handle the demands of simultaneous political and economic reform. For example, countries emerging from conflicts often have unresolved ethnic sensitivities that tend to be exploited in electioneering: hate is an effective campaign tactic. Other states, seeking stability, would be better off providing large-scale employment opportunities through public spending, controlling natural resource assets, and the like until the situation is suitable for the rough-and-tumble world of political and economic reform. Timing is everything, but democratization advocates tend to want it all at once.

There are other ways to have democratic values than the simple formula of elections and market economies and rule of law. We could have democratic governance that favors public control of many national resources, from railroads to mining to health care. We can have democratic process that favors local processes and consensus building rather than national political contests. We can have democracy that flourishes only on the shoulders of a vibrant civil society, and takes heed of the social needs and prerogatives of our most social species. There are many possibilities, and many alternatives examples.

So the United States pursues democratization of a certain, narrow kind. It does so aggressively, and at times with very poor results. We also tend to disparage and even attack democracies that we simply do not like, leading to economic sanctions or coups because we can't stomach the noxious ideologies that democratic governance sometimes puts into power. This is more rather than less likely in the Middle East as it does hold more and freer elections.

One worries most of all that democratization is a stalking horse for other excuses for meddling. It has served the purposes of marketization very neatly. It can also be a cover for American dominance and the imposition of American ideas. The Middle East has been particularly resistant to the kinds of globalization the United States favors. Is it, then, any wonder that it is there that military activism is most feverish?

Democracy is good. Democratization can be and often is, regrettably, not so good. Too bad some people can't tell the difference.

51   Empty (Headed) Threats of War

Among the many foolish proclivities of U.S. foreign policy is the need to threaten "rogue" states and others we dislike when they do or say something at odds with how we see the world. The times this occurs are too many to list, but the targets of threats (and sometimes actions) include a predictable set. Major powers don't really need to do this. A raised eyebrow is usually enough to convey disapproval, or a good old-fashioned scolding like the schoolmarm our presidents tend to imitate.

The problem with threats, especially the empty or hypocritical ones, is that they tend to provoke reactions in the threatened that rebound against us, against civilized values, and often against ordinary people. Three examples are worth brief mention: India, Cuba, and Iran.

The example of India harkens back to 1971. Pakistan, its rival since the partition of British India in 1947, was riven by internal strife, and indeed faced a rebellion in East Pakistan, now Bangladesh. Hundreds of thousands of refugees were fleeing the Pakistanis' brutal repression of the uprising, in which perhaps 500,000 Bangladeshis died, and India was declaring itself partisan on behalf of Bangladesh, finally declaring war on West Pakistan. Conniving as ever, President Nixon and Henry Kissinger—fully aware of Pakistan's atrocities against the Bangladesh people—supported Pakistan's junta, and as a sign of support sent the USS *Enterprise*, a nuclear-armed aircraft carrier, into the Bay of Bengal. This was construed by the Indian elite as an unwarranted and dangerous nuclear threat.

And, by reliable accounts, it stimulated Indira Gandhi and subsequent Indian governments to develop nuclear weapons. "We don't want to be blackmailed and treated as Oriental blackies," said a political party spokesman some years later. The consequences of this "blackmail" are difficult to gauge; India probably would have proceeded with its nuclear program in any case. But perhaps not: without a strategic rationale like the one Nixon served up, domestic opposition to nuclear development would have had more traction.

The case of Cuba is straightforward. Eight presidents have come and gone since Fidel Castro overthrew the tinpot dictator Fulgencio Batista in 1959, and each has threatened Cuba in one way or another, more often by supporting terrorist groups. Kennedy directly threatened invasion. Each threat, implied or indirect or by proxy, has strengthened Castro immeasurably. He can point, accurately, at the threat from the United States to justify his sometimes repressive rule. Without such threats, Castro would have considerably less legitimacy.

Iran is now the flavor of the month, or, truly, the last three decades. But the threat level is reaching all-time heights. Reagan's belligerent behavior included support for Saddam during the brutal Iran-Iraq war from 1980 to 1988 (and never sufficiently protesting Iraq's use of chemical weapons against Iran), and a U.S. gunboat shooting down an Iranian civilian airliner, killing 290 people. Sanctions against Iran have been steady throughout. Even in the reform period of President Khatemi, U.S. behavior toward Iran was generally threatening, and this came to a crescendo in January 2002, while Khatemi still had three years to go in his presidency, when George W. Bush declared Iran a part of the axis of evil.

Few statements could create more havoc for a reformer amid conservative elites than being called evil, and with the United States already in Afghanistan (an excursion with which Iran was cooperative) and about to go into Iraq, the Islamic Republic was certain to react badly. Threats like these strengthen the hard-liners, without a doubt, and the hard-liners in Iran—the Revolutionary Guard, most particularly, that inhaled the chemical weapons and came within miles of overrunning Baghdad—were not about to back down. And, through the ensuing years, the Bushniks have raised the heat higher and higher as Iran insisted on its right to develop nuclear technology. Widespread reports of an impending air strike against nuclear facilities, and even reports of U.S. covert operations active in Iran, helped to usher in the conservative and provocative presidency of Ahmadinejad. This is what's called self-fulfilling prophecy.

When you believe yourself a hammer, everything looks like a nail. The president and much of the ruling elite in this country see the United States as first and foremost a military power, and thus problems are to be solved by armed force, or the threat of it. (A good example in late 2005 was President Bush reacting to alarms about avian flu by saying he would use the military to quarantine the affected areas. Like East Asia, W?)

Who will be next in line for adolescent threats, usually empty, sometimes incompetently carried out? Venezuela is my bet. China might earn a few hollow attempts at intimidation over Taiwan. Perhaps the next country on the U.N. Security Council to demand truth and accountability. The only certainty is that Washington will bellow again and again.

52   America as Victim

Throughout the twentieth century, a segment of the right wing tirelessly insisted that all setbacks to the United States were attributable to internal enemies. America was also beset by external threats, of course, but it was those fifth columnists that really got under the skin. Hence McCarthyism and its many progeny. It is highly septic, this paranoia. Oddly, as America grew ever stronger, outlasted its main rival in the Cold War, and dominated the world economically, this sense of victimization grew stronger.

It has a stout lineage, this particular affliction. In the immediate post-McCarthy period, it had the John Birch Society; the 1964 bestseller *None Dare Call It Treason*, in which Eisenhower was one of the villains; John Wayne, offscreen and sometimes on (*Green Berets*); Young Americans for Freedom; and its apotheoses, Barry Goldwater and Ronald Reagan. Each of these and many, many more were militaristic to a fault, and it is that realm, more than the worship of wealth, sexism, homophobia, and racism that are its domestic underbelly, which is of interest to us here. Betrayal over Vietnam, and then the global betrayal of 9/11, fueled this engine of victimization as nothing before.

That the war in Vietnam can still be described as a betrayal—

"fighting with one hand tied behind our back"—is almost laughable. It was a war in which more than two million U.S. troops were deployed, more bombs dropped than during the Second World War, more than one million Vietnamese killed, and all of Southeast Asia ruined by U.S. intervention. But the sense of betrayal—occasioned by dissent and protest against the war—has outlasted the memory of why America was involved there to begin with. Because American troops became entangled in the politics of the war, there was bitterness and shame that was not as visible after other wars.

That America was somehow a victim of a betrayal (never actually explained) informed domestic politics for years after. Reagan rode this victimhood to victory in 1980, and "standing tall again" was a slogan of his reelection. (That America in the 1970s was also beleaguered by oil dependency, hyperinflation, Watergate, and support for the shah of Iran never seemed to count as much.) When Desert Storm drove Saddam out of Kuwait, the first President Bush said that the specter of Vietnam was buried in the sands of Arabia. And of course John Kerry likely lost the 2004 election because of a concerted campaign to tarnish his Vietnam heroism, because of his efforts in the seventies to end the war.

When the Oklahoma City bombing stirred introspection about domestic terrorism, right wing as it was, an insightful colleague of mine pointed out that Tim McVeigh came from a family that had numerous veterans. It was, he pointed out, striking how the right-wing militias were filled with the poor or lower-middle class Scots-Irish who had derived much of their self-esteem from military service. When that service was disrespected, as many of them believed it was, by the opposition to the Vietnam War, the outrage among some of them was deeply sociopathic.

Any social movement that questioned U.S. policy in the 1980s was painted with the same brush of treason. The antinuclear movement that helped end the wasteful and dangerous Cold

War, the "solidarity" campaigns that opposed the illegal *contra* war against Nicaragua, the antiapartheid activism that eventually led to South African liberation, all of these were blasted as un-American.

These tropes of self-pity continue to this day, and have intensified since 9/11. The shocking atrocities of that day were bad enough, of course. Anti-Americanism generally had roused disproportionate anxiety in this country—how could they hate us when we're so good and generous?—and these attacks seemed to cap that off. It also occasioned new rounds of bloodletting against fellow citizens. Anyone questioning the rightness of bombing Afghanistan or occupying Iraq was called on the carpet of right-wing political correctness and given a good lashing. Not only did Bush's press secretary caution to "be careful what you say," but the columnists began their rhetorical goosestepping: "To march against the war," wrote Michael Kelly in February 2003, "is not to give peace a chance. It is to give tyranny a chance. It is to give the Iraqi nuke a chance. It is to give the next terrorist mass murder a chance. It is to march for the furtherance of evil instead of the vanquishing of evil." And that's the mild stuff.

The victimization was only intensified by how badly things went wrong in Iraq, and, as always, the harpies began to say that critics were undermining troop morale. Reported the *Washington Times* in July 2003: an Arizona Republican congressman said an audio message from Saddam "sounded like some of the statements from opponents of the U.S. leader. . . . 'It sounds like Democratic National Committee talking points. There's just no way of getting around that. It's as if Saddam has picked up on the level of criticism in this country.'" Thirty months later, they were still at it: "Gen. Peter Pace, chairman of the Joint Chiefs of Staff, said during a news conference Thursday that [Representative John] Murtha's remarks about Iraq [that we should pull out] are damaging to troop morale and to the Army's efforts to bring up recruitment numbers." The lack of body armor, grueling rotations, and senseless killing apparently has nothing to do with the morale problem.

The victimization ethos now runs very deep on the right. The popularity of *The Passion of the Christ* owes something to this inchoate sense of betrayal. It accounts for the belief that the Muslim communities in the United States are riddled with al Qaeda cells. It justifies the snooping and torture and renditions and detentions. It divides the country along the brittle old lines of patriotism: the yellow ribbon decals on cars proclaim support for the troops, as if those who do not have them wish the troops harm. It is a diversion from the issue of why troops are there to begin with, and what it would take to keep them out of that region once and for all—a ban, for example, on the SUVs on which most of those decals are placed.

The orgy of self-pity has one definitive impact on the rest of the world—the urge for revenge. The war in Iraq as psychodrama has as much to do with Vietnam as with 9/11 (as it had nothing to do with WMDs). We couldn't get Osama (or Ho), so we whack our old friend Saddam instead. The world complaining about American power? Cut them off financially, ruin the United Nations, and let their diseases fester. Think those Mexicans are stealing our jobs (as maids, car wash jockeys, and busboys)? Build a fence, start a posse, shoot on sight.

Some of it is mere exercise of imperial power, although these sentiments—growing, vitriolic, a mantra of the dominant political party—sound more like the chip-on-the-shoulder attitudes of former empires. That's where we're headed. Much of it comes from the imaginary wounds of the fancied slight. Bad enough to be muscled up and happy to fight; much worse to be a bully with a grudge.

53   The Imperialism of Knowledge

One of the things that happens when one or two countries exert exceptional influence over much of the world is what might be called an imperialism of knowledge—that is, the way that information is shaped, acquired, proliferated, and used tends to reflect the ideas and perspectives of the dominant power. Today, with the United States the hegemonic power in the world, knowledge itself tends to conform to our thinking, like a hot liquid poured into an iron mold. As with most things that stifle diversity, it's not to be welcomed.

The imperialism of knowledge—I use the word *imperialism* to convey some of its unconscious power—is something we scarcely notice here, not only because of American insularity, but because conformity rarely raises an eyebrow. It is dissent that causes the powerful to look twice.

This dominion is never total, another reason why it is not so noticeable. The Soviet Union was never able to achieve this kind of power as completely as it sought to. That is why, after decades of near-total control of the "means of production" of knowledge, the Soviets were vulnerable to cracks in that system—civil society, particularly, which was able to pick up and adapt ideas from elsewhere. But Soviet communism was never a particularly attractive idea; among other things, it didn't create material satisfactions and suffocated the nonmaterial. This was apparent to many, and thus its hegemonic power could be exerted only by force.

But other totalizing systems, if less repressive, were nonetheless quite successful in promoting their ideas. It is much remarked upon that the British Empire left its legal norms and processes, its language, business practices, and the peculiar game of cricket well rooted in the many places it controlled for decades—India and

Pakistan, Australia, much of Africa, and so on. The French similarly exerted such influence in their colonies, perhaps even more so with the implantation of Roman Catholicism, which was also a principal legacy of the Spanish Empire in the Americas and the Philippines.

Religion is a good example of how such imperialism works. It is said that Islam and Christianity succeeded quickly in Africa, Latin America, and parts of Asia because their proselytizers presented them as universally true, superior to local religions; when allied with commercial interests or rival elites, these systems of knowledge gained a powerful place and often triumphed over the local, even where armed force was not used. The same pattern of adopting many other "systems" of knowledge is evident in politics and economics nowadays.

It's not all bad, this imperialism. The domination of a powerful country can bring a number of useful tools of thinking to a place that may not have had them before—science and other forms of learning, for example. After all, the hegemon did not get powerful out of thin air. Knowledge of various kinds typically girds power that is based on more than military brutality alone. Most imperial powers aren't really interested in leaving the artifacts of knowledge production emplaced in the soil of their former colonies. Their imprint on those colonies was instrumental—done to facilitate their own extraction of resources, labor, strategic advantage, and the like, not to improve the lives of the colonized. Even religion was generated locally for purposes of social control more than anything else, at least officially. So we have today, for example, very few good universities in the developing world where France and Britain reigned for so many decades.

America's neocolonialism follows many of the same contours as the British and French. Governments tend to parrot U.S. policy makers in the broad contours of global trends that America dominates—the desirability of market economics, for example, or the war on terrorism. Of course, the economics profession is the principal generator of the knowledge that informs, is reflected in, and reinforces official U.S. policy. The example that springs to mind is the involvement of the "Chicago School"—Milton Friedman and his free market acolytes at the University of Chicago—in reforming the Chilean economy after General Pinochet ousted the leftist president Salvador Allende in 1973. Since then, the market model has become dominant in the world in part because that's what the United States wants and in part because the developing world receives and absorbs the knowledge of the great universities.

And, like most such systems, it tends to be totalizing: not only to create "free" markets, but "free" societies that throw off repressive chains of states, ethnicities, social and cultural backwardness, and so on. In this effort, which is not always a conscious policy of change, the imperial or "universal" vision of how things ought to be ("modernization" is what we used to call it, and now "globalization") tends to trump the local, including what we might call local knowledge.

These indigenous perspectives, habits, customs, and knowledge tend to be portrayed in the West as quaint or repressive. Female genital mutilation is one such practice that is widely decried in the West and probably represents our view of a "local knowledge" that is indeed reprehensible. This is likely an extreme case. Local knowledge can also be about social organization, to cite a more prevalent and important example, which emphasizes sharing large kinship networks that operate in complex ways, stewardship of land and water and other resources, caring for the sick and elderly, and every other aspect of living in a society, however small. The pastoral communities of East Africa provide an illustration. Imperialism enters this arena when development agencies or private corporations or others from outside come in and say, "You cannot sustain (or improve) your livelihood, given the scarcity of land (which is being enclosed by new laws on private property) and water (diverted for or polluted by new industry upstream) or sustain the marketability of the meat you produce (because transnational meat markets are changing the economics). You must abandon pastoral living, settle down, and grow soybean for export." To do so, even successfully (a dim prospect), would demolish everything the pastoralists know about living. That is the imperialism of knowledge at work, every day.

Of course, it works in many other ways too. National elites understand that buttering their bread means reflecting the wishes of the world's major powers. They accept aid, loans, trade agreements, and the many conditions that accompany each, and some of those conditions are about ways of knowing. Television conveys a way of knowing. Books and magazines and the Internet convey ways of knowing. Whoever dominates these media will dominate the ways we see the world, what is considered important, what is considered to be dangerous, how to succeed, and so on. It is a simple and straightforward process, not a conspiracy. What is important, for example, might be conveyed as a particular kind of beauty. What is fulfilling might come via religion, a self-help bromide, a vision of celebrity, or Beverly Hills wealth.

What we impose on much of the rest of the world is not ideology in the sense of the Stalinist dogma that the Soviets failed with, but ideology in the sense of ordering perceptions to conform to certain kinds of categories—freedom, individualism, Christian faith, enterprise, and so on. Not all these are bad. Some of them may be quite liberating and empowering in certain circumstances. What is disturbing is the homogenizing impact, the bland sameness it demands, and that we are imposing these ideas on the world almost unthinkingly, as a by-product of economic power in particular, and we overwhelm the local wisdom and traditions that could enrich us too.

54   Cuba

If there has ever been a more persistently failed policy than the U.S. hostility toward Cuba, it would be hard to find. For nearly a half century, one huffy president after another has tried to assassinate, depose, ridicule, isolate, bully, impoverish, and ban Fidel Castro. Ten presidents has Castro confronted, and will soon have outlasted them.

The history of the relationship is complex, but it boils down to two different perspectives. For Castro and his many admirers in the third world, the United States was a classic neocolonial bully in the Caribbean and Latin America. Among others, we supported the old-style Cuban dictator Fulgencio Batista for twenty years until he was ousted by Castro, Che Guevara, and the broadly popular Cuban revolution in 1959. The United States could not abide by this revolutionary fervor, and had to get rid of Castro as a symbol of independence, as well as satisfy the vengeful rage of the Miami-based Cuban exiles.

For anti-Castro policy makers, the U.S. actions were necessary because of Castro's longtime friendship with the Soviet Union and a brand of Marxism that has been repressive at home and exported abroad. He says he moved toward the Soviets only when he was undermined by the United States.

Defending Castro is not easy. While one could say that he has demonstrated a kind of dedication to his cause that is rare among long-standing ideologues of any kind—he seems neither personally corrupt nor politically expedient—it is difficult to understand his equal dedication to political repression.

Castro remains popular in Cuba, but that sentiment is fading as the generations that recall the hardship of the Batista period are themselves fading away. After the Soviet subsidies ended in the late 1980s, the GNP of Cuba fell by some 40 percent, and this was expected to bring down Castro's regime. But he persevered, and turned around the Cuban economy with the help of outside investment from Europeans. Today, Cuba stands as one of the healthiest of countries in the global south, ranking fifty-second overall in the Human Development Index. Castro's investments in education and health have kept Cuba from spiraling downward even in the worst of times. And it has none of the extreme poverty that beleaguers nearly every developing country and even some of those that rank higher in per capita wealth.

At the same time, there is no political freedom in Cuba. There are forms of democracy, but within tight constraints. Politically based jailings are frequent. Civil society is constantly under siege. One wonders if Castro saw Gorbachev's openness policies in the USSR as a warning that loosening the iron grip is a certain way to lose power.

But whatever Castro's own motivations, the U.S. policy has been a lifeboat for El Jefe. A strange symbiosis has evolved over time. Castro needs American belligerency to justify one-party rule. American politicians need anti-Castro policies to ensure their chances of winning Florida in the next election. Neither is required to do anything inventive or out of character. The U.S. policy now is by rote, and seeks mainly to outlast him (he was born in 1926). Had U.S. political leaders been more flexible, or sensible, they would have opened up with Cuba, just as American policy was open to trade and cultural exchange with the Soviet bloc in Eastern Europe, with good results.

And there's a price for America to pay for all this hostility. It has constantly roiled U.S. relations with Latin America, even those with no taste for Fidel. It elevates him and Che as heroes to millions of young people in the third world. The ongoing attempts to assassinate him and destabilize his regime do not go down well with anyone outside the United States, and tacit or active support for acts of political violence against Cuba rebounds against Washington in the new age of post-9/11 attitudes toward terrorism.

And there Fidel sits, or, more likely, stands at a lectern, revolutionary banners waving behind him, taunting the *yanqui* imperialists, pointing out our maltreatment of the developing world, befriending the leftists of the world: a relentless reminder of how foolish American foreign policy can be.

55   Oceans

. . . the sea is covered with fish which are caught not merely with nets but with baskets, a stone being attached to make the baskets sink with the water. . . .

The great fishing areas of the northwest Atlantic were probably the strongest magnet for European interest in America in the sixteenth and seventeenth centuries, with cod so plentiful they virtually leaped into the boats. John Cabot's famous account (above) was confirmed by generations of fishermen. The richness and plentitude of the sea was a foundation stone of America.

No more, of course. The oceans are fished out, trawled to death. Used as a big trash can. Millions of gallons of petroleum dumped in, toxic chemicals, too. Coastal estuaries overdeveloped, seeping garbage and poisons. Container ships and naval vessels doing heaven-knows-what out there.

Oceans are home to 95 percent of the living space for the earth's plants and animals. Scientists describe a serious decline of ocean habitat, wetlands, marine mammals, fisheries, and coral reefs, while pollution, coastal development, and "dead zones" are increasing at an alarming rate. More than 20,000 acres of coastal wetlands that serve as nurseries for fish disappear every year in America, which is only the tip of the problem for what goes on—largely unseen—farther out in the seas.

The great fisheries have been depleted by overfishing for years, but it's not the fault of the traditional fishermen. It's mainly the corporate fish operations that use bottom trawlers—underwater nets up to one hundred meters wide that are dragged along the ocean floor, damaging or destroying everything in their path. Something like 70 percent of ocean life that is scraped up in the nets is unusable, referred to as "bycatch"—unwanted fish, turtles, seabirds, marine mammals, coral reefs, and sponge forests, amounting to between eighteen and forty metric tons of sea life destroyed *per day*. There are other destructive practices too, like longline fishing, but trawling goes to the bottom where 95 percent of marine life lives.

And there is the trash can dimension: 7 *billion* tons of litter is dumped into the world's oceans each year—60 percent of which is plastics that take ten to twenty years to decompose. Those pretty cruise ships are nasty polluters, too, with 255,000 gallons of wastewater and 30,000 gallons of sewage every day. Once beyond three miles offshore, they're permitted to dump these types of waste into the ocean.

The United States has not been alone in abusing the oceans, of course, but it's a global problem and Washington was for years reluctant to govern the oceans cooperatively with other nations. The inability to stop abuses is clearly a result of this laissez-faire attitude.

But possibly the worst punishment of the oceans comes from emissions that are driving the coming catastrophe of climate change. Almost half the carbon dioxide from the past two centuries of human industry has been absorbed by the world's oceans, an injection of greenhouse gas that could change the acidity of the ocean, threatening the oceans' ecosystems. Levels of carbonic acid—the reaction product of water and carbon dioxide that is found in soda water—are increasing at a rate one hundred times faster than the world has seen for millions of years—drastically affecting the ocean's pH scale, which could alter the food chain and damage coral reefs and sea habitats in dramatic fashion.

Washington has just never gotten a grip on America's destruction of the oceans, perhaps due to the frontier mentality that regards the seas as limitless and bountiful. That they are in fact dynamic ecosystems with complex fragilities and sensitivities is only gradually coming to be appreciated. But the usual band of suspects blocks the more important measures that could protect these fonts of food, oxygen, and beauty.

56   Gangsta Rap and the Culture of Violence

Popular music suffers through cycles of political accusations. I can remember *The Ed Sullivan Show*'s attempt to block Elvis's gyrating body; rock and roll was widely condemned as undermining the morals of youth worldwide. It doesn't take much to undermine the morals of youth, and rock music definitely had that on its agenda. But the morals in question had more to do with laughable conformity to a middle-class ideal, stifling sexuality, obeying our elders, and the like. Come to think of it, rock *was* pretty subversive. But it was not violent, and while it was sexually liberating in certain ways and not in others, it was not misogynistic. And like a lot of American music contributions, it went global very swiftly.

While rock still thrives fifty years since Elvis got all shook up, the one new phenomenon in pop music has been rap, now a $4 billion-a-year industry. Born of the tough neighborhoods of the Bronx in the early 1980s, its rhythmic rhyming verses drawn from early African traditions and street culture, rap quickly ascended and spread, becoming enormously popular in white suburbia as well as its urban milieu. A lot of the early rappers were socially conscious, describing the hardness of ghetto life. An early example is Grandmaster Flash and the Furious Five, whose 1982 song "The Message" went platinum in a month: "I can't take the smell, I can't take the noise / Got no money to move out, I guess I got no choice / Rats in the front room, roaches in the back / Junkie's in the alley with a baseball bat / I tried to get away, but I couldn't get far / Cause the man with the tow-truck repossessed my car."

But before long a new strain, gangsta rap, began to take over. There's been plenty written about it, both outraged condemnations and you-don't-get-it-whitey defensiveness, but the bottom line is a bottom line that is enormously profitable and trading on brutality toward women in particular—pimp culture, treating women like whores, cavalier references to beating them, depicting them in these careless and demeaning ways relentlessly. And then the celebration of a kind of adolescent fantasy of violence, lots of guns and cocaine and revenge and so on, that crosses boundaries of description to adoration. While the music reflects the violence of underclass life to some extent, 80 percent of the buying audience is white, a sign of the power of the thrill of the exotic, a kind of interior orientalism. But whatever its origins and appeal, the messages in the videos and ads and lifestyle and lyrics are undeniably violent and sociopathic.

And, like jazz and blues and rock and roll, gangsta rap *is* embraced around the world. Rap generally is especially popular in Africa, as one might expect, but its gangsta offspring has also found a home in some of the toughest neighborhoods—Liberia, Sierra Leone, Congo, southern Africa—where violence has a much more intensive expression. The teenage soldiers of the warlord armies in West Africa particularly were notably taken with rappers Notorious B.I.G. and Tupac Shakur. As a leading scholar of violence in that region reports, "Calling themselves West Side Boys . . . they incorporated elements of Sierra Leone youth culture into their war fighting culture. The name 'West Side' came from hip-hop music admired by some urban youth. . . . Fighters admired these men who began as enterprising and clever drug dealers, who like them had to live in the informal economy and depend on wits and violence to survive."

The bands of fighters, some of whom practice the most heinous war crimes—forcing children to be soldiers, mass rape, mutilation, and so on—were heard to be chanting the songs of the American gangsta rappers as they preyed on the streets of Monrovia and Freetown. It's a long way from the glam spots of L.A.

As one young friend puts it, "Rap and hip-hop are featured in McDonald's commercials, Reebok and Nike use rap stars to endorse products, movies feature rap artists, clothing companies market hip-hop attire . . . etc. It's everywhere. In the driver's seat are marketing companies and record labels, who use the popularity of the rappers to make gun violence and antisocial, divisive behavior seem not only normal (*not* an aberration), but sexy."

No one could credibly accuse the rappers of stimulating the violence in Africa. But in a world where the number one human rights abuse is violence against women, it's disheartening to know that American lyrics are the backdrop for rape and beatings and murder. Maybe some of Hollywood's celebrity ambassadors could confront their nearby celebrity rappers about this one.

57   Supporting Apartheid

Few insults to human rights rival South Africa's long-standing enslavement of blacks, the system of apartheid. It roiled southern

Africa for decades, introduced personal and communal violence that continues to fester, and impoverished the region. It is regrettable, to say the least, that the U.S. government was so slow to oppose it.

The tiny white minority of descendants of British and Dutch colonizers created a Nazi-like political and social system of confining native African blacks to shantytowns—deplorable living conditions—and denying them all but menial jobs. Apartheid similarly held down South Asians who had migrated to South Africa. (It was his experience with racial injustice that made an activist of one such migrant, Mohandas Gandhi.) Until popular movements and global opprobrium finally forced the whites to cede power in 1989, South Africa was a pariah state, aggressively attacking its neighbors, developing nuclear weapons, and committing heinous crimes against tens of millions of blacks. And through it all, the United States essentially looked the other way.

That is, the U.S. government and its right-wing establishment looked the other way. The left organized boycotts, divestment, sanctions, campus activism, and rallied world opinion to the side of Nelson Mandela and his cohort.

But official Washington consistently balked at any such action to end apartheid. When an arms embargo was imposed on South Africa in the early 1960s, the United States found loopholes. When nuclear weapons development was discovered and publicized by the Soviets, Washington stepped up to help Pretoria. When the world community sought severe economic sanctions and support for African sovereignty, the United States sought a policy of "constructive engagement."

For example, years after the nuclear weapons program was discovered by Moscow and confirmed by U.S. intelligence, U.S. secretary of state Alexander Haig and South African foreign minister Pik Botha met in May 1981. As the National Security Archive recounts it, "Perhaps as a result of that meeting, the

Reagan administration increased nuclear-related assistance to Pretoria by approving exports of nuclear material, computers and high technology items to South Africa. The administration also renegotiated with Pretoria for the resumption of uranium imports, prohibited when Congress passed the Nuclear Non-Proliferation Act in 1978."

The reasons for looking the other way were not only the apartheid regime's stout anticommunism, but the country's vast mineral wealth. Moral blindness comes easily to Washington whenever money is at stake. A touch of racism cannot be discounted, either.

"Throughout his tenure," says the Archive, "President Reagan studiously avoided criticizing the South African government, repeatedly praising the Botha Administration for making substantial reforms despite the overwhelming evidence of the continued and extensive exploitation and oppression of the black majority in South Africa. He has directly and openly embraced the Botha Administration as 'an ally and friend,' demonstrating what critics saw as a callous indifference to world-wide demands for human rights and basic freedoms for the blacks."

Reagan's policy of constructive engagement was undermined by Pretoria's belligerency in southern Africa and by the wars, often supported by Reagan, which took so many lives in Angola and Mozambique particularly. South Africa fought mightily to contain or reverse the progress made in the former Portuguese colonies, liberated by Marxist movements, and in Namibia, which was occupied by South Africans before they were ousted by a popular movement there.

Along the way, the essential American support for Pretoria's policies, punctuated by dissembling that responded to almost universal condemnation of U.S. complicity, delayed popular sovereignty in southern Africa and crippled the effort to restore order and promote prosperity in those countries. They still feel the effects of those pro-apartheid years and the insensitivity of economic and health policies since.

Consider the testimony of Michael Lapsley, a priest, incarcerated during apartheid.

Vast numbers of people were imprisoned. It was during those years . . . that torture became normative. It became a principal weapon used by the Apartheid regime against people, particularly against black children during that period. It was also a period where there were a vast number of people on death row in South Africa. Every Thursday, up to seven people at a time were executed, but it was also a time when the Apartheid regime was in the rampage in the Front Line States attacking Botswana, Lesotho, Mozambique and Zimbabwe. There were a number of massacres of refugees that took place. It was also a time of civil war in Angola. And it was the Reagan administration that was supporting the Unita bandits in Angola and fomenting war. And it was clear to the people of South Africa during those years, that whilst there were a vast number of ordinary people in the United States, particularly African-Americans who stood with us, the Reagan administration was on the side of Apartheid. It was both Reagan and Thatcher who were giving succor to the Apartheid regime and in a sense prolonging our struggle. More people had to die in South Africa because of the support that came from western governments, particularly from Washington and London at that period.

"Prolonging our struggle." That is the simple and straightforward truth of the matter—Reagan stood with the racist and repressive regime, braced by his right-wing base at home, mindful of the massacres and nuclear weapons and harsh imprisonments and the warfare in the neighboring states. Supporting the bad guys every step of the way. And now, years later, we say, "Look what a mess Africa is. Why can't they pull themselves together?"

58   The Very Expensive Fissionable Atom

After the United States attacked Japan with nuclear weapons in 1945, some public concern began to grow about the destructiveness of the weapons. To counter that, President Eisenhower in 1953 initiated the "Atoms for Peace" program, promising all in the "free world" nuclear-generated electricity. Now, 442 nuclear power plants are in operation worldwide.

With climate change a clear and present danger to Earth's ecosystems, the nuclear power industry seems to have new life. It had suffered widespread doubts brought on by the technology's high costs, waste disposal problems, and safety, particularly after the devastating 1986 accident at the Chernobyl nuclear plant in Ukraine and the 1979 accident at Three Mile Island, Pennsylvania. For a time, mainly in the 1970s and 1980s, sizable social protests against the nuclear power option seemed to signal its decline. No new nuclear power plants have been started in the United States since the mid-1970s. Nuclear weapons proliferation also goes hand in hand with nuclear power development, as India, Pakistan, Israel, and a number of wannabes have shown.

Regrettably, the industry has failed to solve some of these pressing problems. Life cycle costs of nuclear power—including R&D, government subsidies, and the still vexing matter of waste—are higher than alternatives, including burning fossil fuels in more green-friendly ways. And the idea that we can somehow save ourselves from climate change by building nuclear plants distracts us from more sensible alternatives. (It may not even be true that nuclear plants are better for climate change; as the quality of uranium ore deteriorates, a nuclear economy may be worse for the atmosphere.)

Those alternatives include energy efficiency—vastly improving the technologies we use, like refrigerators and air conditioners—and conservation, simply using less. The "using less" part seems distinctly un-American (the consumer ethos we've cultivated is seemingly insatiable, and it soaks up energy), but it's been done before when need be. Efficiency and conservation alone, if given the priority in government and industry that nuclear energy and petroleum have enjoyed, would provide a clean bridge for a decade or more. (Nuclear energy has benefited from enormous government subsidies, something on the order of $100 billion, with many expensive issues like waste and aging plants not solved. There is less than $1 billion for all renewable fuels and efficiency in the federal budget.) At the other end of that bridge are other renewable fuels, now gaining a new life as the price of oil rises. Among those new fuels, conceivably, could be a much better nuclear technology, most likely fusion, which does not produce waste or material for bombs.

On balance, it would have been preferable to do without nuclear energy altogether. It has added relatively little to the world's energy supplies—about 7 percent—and has exacted an enormous price in return. Without it, we would have had to confront our energy demons—consumption, uncertain oil, climate change—sooner and perhaps more effectively. The possibilities for nuclear weapons proliferation would have been reduced to near zero, along with the military costs associated with that problem. It's easy to say now that nuclear power was a costly gamble gone awry, and had we invested sensibly in alternatives our choices would be better today. This is only too obvious, and legions of scientists and activists were saying it beginning more than thirty years ago. It remains a valid point today, given nuclear power's high costs, uncertainty, and dangers.

It was a nice dream, the "atoms for peace," and might have been a great gift to the world. But not all dreams come true, and this one certainly did not.

59   AIPAC

The American-Israel Public Affairs Committee, or AIPAC, is the apotheosis of the lobbying group as a representative of a foreign country. Few lobbying organizations have been more powerful. Few have more single-mindedly pursued a narrow agenda in U.S. foreign policy, and few have done more damage.

AIPAC is largely Likudnik in outlook, which is to say, the right wing of the Israeli political spectrum, and it has, through campaign contributions and the other mechanisms of our gilded democracy, been able to push American foreign policy in that direction as well. (Many say it supports whatever government is in power in Israel, which may be accurate, but the overall impact of its lobbying is more conservative.) Many American Jews are pro-Labor and contribute substantially to NGOs that promote peace and reconciliation in the Middle East, so one cannot simply call AIPAC the "Jewish lobby" and leave it at that. It reflects the caution and even militancy that many diasporas tend to embrace. But even with understandable concern about the viability of the state of Israel, many American Jews support the peace process and justice for Palestinians.

The activity of AIPAC and its think tank, the Washington Institute for Near East Policy, tends to the more belligerent side of the equation. The Institute has organized a number of study groups and commissions that had exceptional influence on U.S. policy, in part by recruiting those who are likely to enter government. The drumbeat on Iraq and Iran particularly has provided important support for the official U.S. hostility to those regimes, the results of which have been catastrophic. (While a majority of American Jews opposed the war against Iraq, AIPAC insisted on it.)

The impact on the Israeli-Palestinian issue is troubling enough in itself. A well-financed lobby at over $30 million a year, AIPAC can get a lot of bang for the buck, and it is relentless in rooting out critics of Israel and congressional action that might displease the Israeli government. The Center for Responsive Politics calculates AIPAC's contributions to candidates of both parties at greater than $20 million between 1990 and 2004. For this it can get sanctions on countries it believes are a threat to Israel, can block any attempt to constrain Israel's more destructive policies—treatment of Palestinians generally, building settlements, obstreperousness on the final status of Jerusalem—and has helped engineer numerous military assistance deals to Israel.

The upshot is U.S. policy that is less flexible than it should be, and the Arab world sees this as an attack on its interests. This can and does lead to more support for militancy among Palestinians. The sanctions against Iran, even in its long reformist phase under President Khatemi, gave succor to the right-wing mullahs who have now regained a stranglehold on Iran's politics. And the enormous levels of financial aid to Israel (and, by necessity, to Egypt) have crowded out more sensible choices.

All of this is said by one who is generally pro-Israel. It is a democracy, very often a social democracy, and its creation was sanctioned by the United Nations. But the influence of essentially foreign agents on U.S. policy making, and the acceptance of its vassal think tank as an independent analyst on international relations, are dispiriting failures of the policy and opinion elite in Washington, and they have obviously deleterious consequences. At the same time, liberal Jews in the United States have failed to create an alternative to AIPAC that speaks for them.

I would not go so far as to say that the U.S. position on Israel is a cause of Islamic terrorism across the globe, but it does not help. I also reject the notion that the United States has some moral obligation to support the Palestinian cause, although the correctness of statehood has long appeared unassailable. The frustration of many Palestinian activists is a result of the unequal treatment they receive in the news media and in political discourse here, and how they are left out of Washington decision making. That is partially a consequence of AIPAC's lobbying. If one is serious about bringing peace and democracy to the Middle East, then rejecting the narrow interests of AIPAC would be an excellent place to start.

60   Wars of Choice

The history of warfare is a history of choices, usually of political and military leaders who intentionally go to war to save their own hide rather than to defend a country or a principle. Somebody always has to go first, and it was not always a madman like Hitler.

So the notion that President Bush has constructed a new doctrine of preemptive war, or a "war of choice," is a little misleading. Intentionality is always at work, and alternatives are almost always available.

Even some of the most justifiable wars look, in retrospect, avoidable. The three great wars of our own history can be viewed in this way. The Revolutionary War was supported by a mere third of the colonists, was fought in part to benefit the wealthy classes in the colonies, and, given the way places like Canada and Australia evolved, was hardly a matter of throwing off the chains of enslavement. The Civil War was possibly avoidable if slavery had been abolished earlier (as it was in all "civilized" countries long before our own; for example, in all of the British Empire by 1833)?

or made a part of our political foundation, which after all had extolled the equality of all men (and, later, women). But even were the South to secede as it did in 1861, what would have happened had Lincoln let it go? Slavery would have persisted for a few years longer, but not much more, and the Confederacy would have been a small and isolated country, probably reuniting with the Union some years later.

Averting the Second World War is a harder case to make, but it is now widely recognized that the punitive peace imposed on Germany after the First World War directly led to Hitler's rise. That is why America took a much more comprehensive view of its global role in the aftermath of the Second World War, creating the United Nations and other institutions to stabilize the world.

The current wars of choice were avoidable as well, and it's not just in hindsight that we can see that. The Reagan Doctrine at work in Afghanistan promoted a violent mujahideen to harass the Soviets in the late 1980s, which came home to roost with bin Laden and the Taliban. The neglect of Afghanistan after 1989 resulted in a festering civil war that was most hospitable to the likes of al Qaeda.

The longtime flirtation with Saddam Hussein was similarly avoidable, built on the mistakes made in Iran and our strategy of balancing against the mullahs who came to power there in 1979. Even after the debacle of supporting Saddam during the 1980–88 Iran-Iraq war, the Bush One crew should have recognized him for what he was (they already knew of his threats against Israel and the use of chemical weapons against Kurds and Iranians), and drawn the famous line in the sand long before he occupied Kuwait in 1990. Turning the screws on him long ago (which might, in a truly enlightened policy, have included lessening our oil dependency) may have given us many more options short of war in 1991, and certainly in 2003.

So the fuss about the wars of choice that the current President Bush has pursued is a little disingenuous. Wars chosen by great powers are almost always avoidable. That also means that the doctrine of preemption Bush promulgated in 2002 is both in practice an empty gesture of belligerence and an unnecessary insult to international law and common sense. But we've come to expect no less of him.

War, we are often told, is a last resort. But for presidents, whose constitutional power is in part built on the role of commander in chief of the armed forces, it is often a choice made to pursue politics by other means. War nourishes and enhances presidential power. It always has. It has an almost irresistible allure, especially for those who are faltering as a result of other political shortcomings. The impact on the rest of the world for this constitutional imperfection is only too obvious.

61 Haiti
Voodoo Foreign Policy

Like Jamaica and a few other Caribbean countries, Haiti has gotten a little too much attention from the United States sometimes, and neglect at others, and neither has worked out well for the Haitians. At the time of President Clinton's intervention to restore Jean-Bertrand Aristide to the Haitian presidency in 1994, Senator John McCain of Arizona said that we owe nothing to Haiti, we have no responsibility that should draw us in. It's a commonplace view, and dead wrong. It has not kept Washington from meddling in Haiti's political process, either, which it continues to do—rendering Haiti another victim of the U.S. version of democratization.

The United States occupied Haiti for nineteen years beginning in 1915, in an "invited" arrangement to restore order after a long period of unrest, but as usual with these episodes, the order was meant to favor certain classes over others. The United States wanted to establish commercial and strategic stability, and occupation and nation building were viewed as our right. The occupation was resisted by sizable rebellions and "scores" of Haitians were killed on several occasions. "The order, however, was imposed largely by white foreigners with deep-seated racial prejudices and a disdain for the notion of self-determination by inhabitants of less-developed nations," reports a Congressional Research Service brief on the country. "The whites from North America . . . did not distinguish among Haitians, regardless of their skin tone, level of education, or sophistication. This intolerance caused indignation, resentment, and eventually a racial pride that was reflected in the work of a new generation of Haitian historians, ethnologists, writers, artists, and others, many of whom later became active in politics and government."

The military was professionalized and assumed power upon the U.S. departure. "As in other countries occupied by the United States in the early twentieth century, the local military was often the only cohesive and effective institution left in the wake of withdrawal." This led to a series of repressive dictatorships, punctuated by the leadership of a populist Dumarsais Estimé, a former school teacher, who came to power in 1946, but was ousted by the mulatto elite when his policies appeared too left leaning. The disarray that his ouster ushered in led finally to the triumph in 1957 of François "Papa Doc" Duvalier, who at first appeared as a reformer but soon betrayed that for total dictatorial power. He defanged the professional military in favor of personal loyalty and set up rival militias, including the infamous *Tonton Macoute*. He remained in power until his death in 1971, whereupon his son, Jean-Claude ("Baby Doc"), took power. He continued the corrupt and brutal rule that sent tens of thousands into exile and murdered tens of thousands in Haiti during his reign of terror.

Throughout this period, the United States stood by and tolerated the bloody antics of the Duvaliers. Both Eisenhower and Kennedy sent aid to Haiti that was diverted to strengthen the terror apparatus. The rationale for tolerating the repressive Duvaliers was Haiti's strategic importance in relation to Cuba: the U.S. government consistently preferred a corrupt thug to the possibility of another populist coming to power in the Caribbean. That is precisely what happened when Jean-Bertrand Aristide, a priest to the poor, was overwhelmingly elected in Haiti's only democratic election in 1990. He was ousted by a military coup the following year. His "restoration" was short-lived and tightly constrained, and the United States persisted in blocking populist politics—and international aid—consistently scheming to undermine Aristide in particular. Aristide's most recent ouster in 2004 was probably engineered by internal "irregular" forces—one could call them terrorists—trained and supported by Washington via the International Republican Institute. It just never ends.

The sad story of Haiti—violent, impoverished, and ravaged by AIDS—is not one that can be placed at America's door. But our responsibility is apparent, and in some ways it is typical of how the United States acts toward the poor countries of the world where we have "interests" that conflict with the simple desire for peace and economic equity. From occupations to misplaced aid to hostile neglect and outright destabilization, the United States treats these kinds of countries as discardable pawns or nuisances.

Most particularly, Senator McCain and others of like mind illustrate how ignorant Americans tend to be of our own history of meddling, misbegotten or malicious policies, and all. It is easy for them to then overlook continuing meddling and disruption fostered in the halls of Washington power. To say the United States has no responsibility for Haiti is like saying we have no responsibility, period. Moral bankruptcy could not find a more vivid example.

62   McDonaldization

Who does not salivate at the scent of McDonald's french fries? Who has not been grateful for a Whopper when the alternatives in an airport or roadside plaza were so bleak? We're all fast foodies in one way or another. What is Starbucks if not fast caffeine and pastries? Isn't the mom and pop stand at the beach just a smaller version of the same? They come in many shapes and sizes. Some, like White Castle, are celebrated as nostalgia-inducing relics.

But of course there is much amiss in the realm of Mickie D's and its many imitators. Not with its growth or profits, but with the influence they have wrought. Fast food may be the quintessential expression of Americanness abroad. It is surely the most pervasive contribution we have made to global cuisine. And while it is easy to be snobbish about it, and lament its astonishing acceptance worldwide—from Paris to Phnom Penh—it is undeniably a phenomenon of the highest commercial order.

Perhaps nothing has brought American consumer culture to the rest of the world as quickly and ubiquitously as McDonald's et alia. Of course, the story begins at home, where our compatriots spend $120 billion a year on fast food, more, says the definitive *Fast Food Nation,* "than on higher education, PCs, computer software or new cars, or on magazines, going to see films, recorded music, newspapers, videos and books combined." The health impact of eating these high-fat, high-sodium, high-sugar foods would be difficult to calculate.

Americans are the fattest people on earth. A major academic study in 2004 says that "since 1980, obesity rates in the United States have increased by more than 60 percent in adults, while rates have doubled in children, and tripled in adolescents," and the average family's total caloric intake from fast food over the previous three decades rose by four to five times. "In 1970, Americans were spending $6 billion on fast food, and by the year 2000, Americans were spending $110 billion."

The obesity problem in children is acute, described by many health professionals as an epidemic. Children who eat fast food add weight, an average of six pounds annually. When poor exercise and TV-viewing habits are added in, a major problem of health and quality of life is apparent. Not to mention health care costs. As these obese children grow into adulthood—and more from their generation become grossly overweight—the numbers afflicted with heart disease, colon cancer, diabetes, and other diseases will increase. The children are a major target of MacDonald's advertising, and its on-site playgrounds are a big lure.

It doesn't stop there. The litter from fast food emporiums is a major problem. "Ever since McDonald's, KFC, and the other big fast food brands arrived in New Zealand our roadside rubbish problem has grown," a Green Party spokesperson claims. "It is a standing joke amongst the fast food operators that, as long as it is dropped out of sight of the outlet, they regard their litter as free advertising."

The industry has lobbied hard to block worker safety and minimum wage advances. This runs the gamut from the slaughterhouse to the food counter (McDonald's is the largest beef merchant in the world). It has quashed hundreds of attempts to unionize.

These qualities are now spreading like cancer around the world. McDonald's, which derives most of its profits outside the United States, has more than 30,000 restaurants in some 120 countries. In China, fast-food consumption is booming. Importation of french fries from the United States increased tenfold from 1995 (there's a solution to our trade deficit!) while the obesity of teenage Chinese children has tripled. KFC is bringing its special brand of high-fat chicken 'n' fixins to China at an aggressive rate, and around the world. The Colonel's secret recipe is available in 90 countries.

Now, of course, one can say this is all voluntary. But the U.S. government is helping out, too. There are not only the usual giveaways to corporate agribusiness, diplomatic and military muscle applied on behalf of trade, and so on, but direct subsidies for marketing fast food abroad, nearly $100 million a year.

The confluence of different threads of American dominance in the world sometimes combines in unexpected, even bizarre ways. Consider this news report from March 2005: "Karachi — Six people were burnt alive when a mob protesting a suicide bombing of a mosque torched an outlet of an American fast food chain in the southern Pakistani city of Karachi, police said on Tuesday. Police and firemen recovered the bodies of six Kentucky Fried Chicken (KFC) employees after an angry mob set the restaurant on fire late on Monday following a suicide attack on a Shiite Muslim mosque here that left five people dead."

Those six, sadly, died spectacularly. Many more millions will simply die early from heart disease and strokes. We're projecting our unsavory habits very deftly, from fast food nation to fast food world.

63   Dissing the United Nations

It is easy to make fun of the United Nations, with its earnest "one world after all" patina and scurrying bureaucrats and endless declarations of goodwill and concern. As an occasional venue for criticism of the United States, it has earned the everlasting enmity of the American right wing, a level of hatred and vehemence that was once reserved for Stalinists. The rhetorical battering of the citadel of multilateralism on the East River is so routine now that in ordinary political discussions in America there are very few defenders of the United Nations. The pity is, the United Nations is actually the invention of the United States, and we have happily utilized it to advance our interests for sixty years. Undermining it is bad for us and bad for everyone else too.

To remind us all, the United Nations and its many off-shoots—the World Bank, the International Monetary Fund, the World Health Organization, the World Food Program, UNICEF, etc.—were created immediately after the Second World War with the United States as its principal sponsor. The U.S. action, which was immensely popular for years, was driven by a certain idealism, to be sure, but also by the recognition that great powers require multilateral institutions—treaties, organizations, principles, etc.—to stabilize the world. Otherwise, it falls to these great powers to do the stabilizing, which, in the aftermath of two world wars, looked a bit daunting. The idealism came from a new commitment to popular sovereignty and the end of colonialism. Providing the new and weak nations of the earth a voice in global affairs seemed not only sensible, because it kept them in the fold of our ideas and interests, but fair. And so it was on both counts.

This vision has served the United States, and the goals of the United Nations itself, quite well, if imperfectly. It has provided many services to the developing world that would have been difficult to manage without a world body. It has given the poor and powerless a voice, not an inconsiderable thing. That the General Assembly in particular has witnessed many an anti-American diatribe should tell us more about the grievances at work in the world than anger us for their seemingly bad manners.

Yet this is what gets the goat of so many right wingers, who have always opposed "world government" and were determined to shoot it down from the start. It is as though no questioning of America's motives or actions is allowable, not in a house we built! And the corruption! Yes, well, the United Nations has often been its own worst enemy when it comes to its management, but perhaps no more than any enormous organization. Corruption was not invented in the secretary general's suite.

The right wing diatribes ignorantly mistake form for substance. The latter is in the trade agreements, rules of international law on numerous economic matters, and development assistance that often favor U.S. corporations above all else, and have been eagerly sought by all presidents and congresses. The provisions of disease control, food security, education for children, and so on are a godsend, and they would not have been likely without the United Nations. The instability that the lack of these services might have wrought is chilling to contemplate.

But most tellingly, especially in the post-9/11 climate, we can see that the United Nations is a place to vent the frustrations that developing countries build up over the years. It is, in this way, a useful place to learn, to reply, and to respond constructively. How else do we learn? How else do they learn? The United Nations can be and is a place of relationship building, problem solving, and cooperation like no other.

By disrespecting the voices of the small and weak, or the contrarian, or the occasionally anti-American voices, we are in effect saying that a civil, largely democratic, open forum *that we created* is a tool of the dismissible and disgruntled, and we're not going to listen. We're not going to pay our dues. We're not giving you any standing in the world or in America itself. Go away.

And where do they go away to?

64 Mel Gibson

Perhaps above all his other qualities—his rugged good looks, athleticism, B-quality acting—Mel Gibson most exudes one thing that's quintessentially Hollywood-American: violence. When it was just dumb cop and revenge flicks, it was forgettable. But then there was *The Passion of the Christ,* the apotheosis of sociopathic violence masquerading as morality play.

Much has been written about the 2004 film's anti-Semitism (which is unmistakable) and Gibson's promotional gimmicks (which worked like a charm). But it's the violence that keeps the film humming. It's the blood and guts that keep the believers agog and filled with joy, with pastors renting out cineplex theaters to show the unrelenting violence to teenagers. "It is as it was," Pope John Paul II reportedly said after a special viewing at the Vatican. Jesus died for our sins, Gibson reminds us, and we should witness this death because we are all responsible for his suffering (a concept that remains utterly obtuse to me fifty years after first being exposed to it as an altar boy). Gibson's martyrdom runs through most of his films, soaked in blood; that he should take up the ultimate martyr's case is unsurprising. The violence *is* gruesome. Forty-five minutes of lashing, cuts, and of course those nasty nails. It can't fail to make an impression, as indeed the story has for two millennia.

Two striking aspects of the film are noteworthy. The first is how Christian admirers of *The Passion* seem to feel vindicated that all this suffering somehow proves Jesus's uniqueness. But of course many thousands of political prisoners have endured years of torture, flagellation, and horrid (and anonymous) deaths. Many of these have been tortured and murdered by American Christians, or other Christians. (A disproportionate number of the victims have been Jews, including Jesus.) None of this brutality—far worse than what Jesus experienced—is taken as a sign of divinity. Indeed, the whole topic of political prisoners is now relegated to the back pages of our global narrative.

Even within this lexicon of Christianity, *The Passion* conveys the wrong message of Jesus, which above all else was one of forgiveness and redemption. As the Catholic intellectual James

Carroll puts it, there is no resurrection in the film, which is the central mystery and power of Jesus. "Not the Risen Jesus, but the Survivor Jesus," Carroll explains. "Gibson's violence fantasies, as ingenious as perverse, are, at bottom, a fantasy of infinite male toughness."

Why does this matter? This most visible film, watched in the White House and the Vatican, stirring controversy a year before opening, then grossing the most money ever by an independent film ($700 million worldwide in five months of 2004), sends a very powerful message around the world. We, the most Christian of nations, proclaim that our messiah's life and meaning is most vividly understood as one of violence. This is a guiding moral vision, moreover, of a people known to suffer the lash of Semitic terrorism.

The message has been received. The film has drawn tens of millions of viewers around the world. Hollywood has green-lighted new pseudoreligious projects to capitalize on Mel's pioneering trail. America has now done for violence what no one has done before, defining it as the essence of the religious experience, now available on DVD in glorious Technicolor.

65   The *New York Times* (and the *Washington Post*)

In almost every way, the *New York Times* is the best daily newspaper in the world. In its reporting, the *Washington Post* is not far behind. Both aspire to excellence in journalism—getting the story, giving it context, providing depth, linking it to larger issues, following up. They may not always be ahead of the curve, but when they hit the bend, they're moving fast and sure.

Except when the terrain is U.S. foreign policy, where both frequently falter. They don't measure up to their own standards or the expectations of many readers, in part because they have a rigid view of the world on some pivotal issues that is unshaken by fact or fancy. This leads both, in different ways, to poor judgment and large blind spots in the understanding they bring to their pages. Two matters above all else reflect these ideés fixes: globalization and war.

To the editorial writers and news editors, globalization above all else is about free trade in its several forms: not merely low or no tariffs, but unencumbered flows of capital, protected intellectual property rights, freedom of foreign ownership, and so on. There are other forms of globalization, of course, including the cultural and political. But those are treated as either the developing world's unrequited love for American pop culture, or as a hunger for democracy. A more superficial reading of these phenomena would be hard to find, but there it is.

On trade, however, the coverage is ample and serious, on business pages, of course, but as a news emphasis, too. The possibility that these forms of "free trade" may not be the best options for people in the developing world, or for the environment or labor standards or stability, is not what earns attention. The unbreakable consensus of the power elite in the United States is that globalization is good, free trade is particularly good, "raising all boats," to use the hackneyed image, and certainly a boon for America. The picture is much more mixed than that, as levelheaded economists will admit, but the *Times* and the *Post* see the conflicts in globalization as a simple matter of politics: the protectionist unions and antieverything protesters against the wretched of the earth. The antiglobalization forces are blocking Progress and Good from being brought to the world's masses, as always, by a selfless and beneficent America and the invisible hand of the market.

The coverage and the editorials are more nuanced than that—usually—but the central message is to support "liberalized" trade above all else, and to belittle those who raise doubts.

The wars of the post-9/11 period, and even in the post–Cold War era, show a similar tendency to undergird American power and perspectives. All the military actions of the Bush administration have been endorsed, often heartily, to the point of near fabrication of stories or willing support of the White House's own fabrications. The *Times* at least has taken some steps toward self-examination in this regard. The *Post*, especially in its editorials, still cannot acknowledge that the phony reason for going to war in Iraq requires correction or apology. The extent to which error is acknowledged typically comes out as "everybody thought they had WMDs then," which is demonstrably untrue.

But it is not just the errors of a star reporter like Judith Miller of the *Times* or a clueless editorial-page editor like Fred Hiatt of the *Post*. The problem is the more penetrating and durable bias and predilections, as in, for example, the apparent editorial policies not to delve into the matter of Iraqi civilian casualties with the same interest shown in, say, the new fashion sensibilities of Afghan women. The *Times* went so far as to report a previously unheard-of Iraqi think tank saying that, as of late 2005, there had been 500 civilian deaths attributable to coalition forces in Iraq, which is low by a factor of about 200. That it could say so without a shred of awareness of the importance of such shoddy reporting is an indication that its mind was made up—namely, there would be no actual reporting on this issue because, in all likelihood, it's too sensitive.

Or consider the war "within." The terrorist threat to the United States has made good copy since the news media switched its focus from a missing congressional intern to Osama bin Laden in the late summer of 2001. The newspapers report, quite routinely now, the many indictments and investigations of mostly

Muslim men in the United States who are charged with this or that "terrorism-related" crime. When someone is acquitted or pleads down to a much lesser (and nonterror) offense, it gets far less if any coverage. But those cases are in some sense at least real, which is to say, real arrests and prosecutions, however trumped up or overstated.

More odious are the implications of a threat without a scintilla of evidence. In August 2004, an old intelligence report, never actually verified, held that al Qaeda operatives were casing buildings in the financial district of New York and the World Bank and IMF buildings in Washington. The headlines were grand, with photos of Special Forces guarding the buildings. In the body of the *Times* story, it was strongly implied that there were domestic jihadist operatives. That there is no evidence of this, and that it only reinforced the unwarranted fears that conveniently earned George W. Bush a second term three months later, was never part of the story in the *Times* or the *Post*. (Although, to its credit, the *Times* has done one of the better pieces on this topic, covering the dubious prosecution of the "Lackawanna Six" in a thorough story by Lowell Bergman.)

What is especially important about these two newspapers is that they shape the coverage of these issues for the entire American news media, and some foreign news organizations as well. Politicians get most of their information from them. Years ago, having done work on strategic defense and intrigued by how Congress absorbs such technically complex information, I commissioned a survey of how members of Congress get their information on defense issues. We found that studies, hearings, policy journals, and the like scored very low. By far the overwhelming sources were the *New York Times* and the *Washington Post*. So our behavior as a nation—official policy, openness to dissent, perceptions of those in the rest of the world—is largely shaped by a few editors and the owners at these two newspapers.

An ideology is a way of constructing reality and rationales to match one's class or sectarian bias. Journalism prides itself on being above ideology, and indeed loves to make fun of ideologues. There are few sturdier ideologies than the Washington consensus on free trade, or the suspicions of the loyalties and honesty of Muslims in foreign capitals or nearby American neighborhoods. That two of the very best news organizations in the world cannot shed this ideological lens when reporting and commenting on the news is astonishing. More important, it has a powerful effect on how all American journalism reports on these subjects, and how policy makers receive and act upon information. And that is a can't-miss recipe for screwing up the world.

66   Blame It on Rio
The NRA's Shoot-out in Brazil

Brazil has the highest rate of gun violence in the world. Its reform government passed gun control legislation in 2003 that reduced the fatalities from handguns dramatically, but the problem persists. In 2005, a referendum to ban the sale of guns to individuals appeared to be on its way to an overwhelming victory, with national polls showing four-to-one approval. That's when the National Rifle Association, the odious American gun-loving powerhouse, swung into action.

No good trend could be allowed to gain momentum, so the NRA advised the local gun owners group on how to beat the ban. One correspondent described the resulting television ad campaign as especially effective:

> The ad starts with a sober, simulated news report. A news anchor, looking directly into the camera, warns viewers

about Brazil's proposed gun ban. "People are misrepresenting the disarmament issue," she says. "It won't disarm criminals." The anchor fades and a news-on-the-march montage begins, highlighting freedom's red-letter days. Nelson Mandela is released from prison. A single man impedes a row of tanks in Tiananmen Square. The Berlin Wall falls. "Your rights are at risk," says the anchor, returning after the inspiring film clips. "Don't lose your grip on liberty." And then, to bring the message home, archival footage runs of thousands of Brazilians taking to the streets, restoring popular rule after more than two decades of dictatorship.

The gun ban was defeated. The NRA crowed that they would do the same anywhere that "gun rights" were threatened. Taking the American credo of "more guns, less crime" to the rest of the world is a harrowing prospect. (Brazil proves that more guns equal more crime.) And the NRA not only is vigilant about defeating any reform in any corner of the globe, it protects the "right" to sell small arms pretty much to anyone, anywhere, anytime.

Some 600 million of these weapons are in circulation globally, and of the forty-nine conflicts, mostly civil wars, in the last decade, forty-seven of them were mainly fought with small arms and light weapons.

A major source of supply is the four thousand gun shows in America, where one can buy a startling array of weaponry. At one in Kentucky, according to an exporter, "table after table was laden with AK-47s, M-16s; sniper rifles; handguns; armor-piercing, tracer, incendiary, flat, and round-tipped bullets; silencers; night scopes; Civil War, World War I, and World War II weapons; Nazi paraphernalia; and Japanese swords. There were even a couple of antiaircraft guns." Some of these are rather powerful weapons, like the Barrett .50 caliber rifle that can blow away a target at two thousand yards. They can be legally exported by individuals.

And they are, to Liberia and Sudan and Kosovo and Congo and Pakistan—and Iraq. Where are the insurgents in Iraq getting their weapons, estimated to be at least 7 to 8 million small arms? Well, they could come from many sources, but the fact that they could remain in circulation was a policy of the occupying power. "When the United States turned over sovereignty to the new government of Iraq," wrote arms expert Rachel Stohl in 2004, "it did so without confronting one of the most pressing problems facing the country: the millions of small arms and light weapons plaguing Iraq's security and threatening its stability. Excluding small arms from the long-term security plan is a deadly mistake." I guess so. But it's consistent with American gun fetishism.

And the new attempts to control this river of weapons have been stoutly undermined by the NRA, first and foremost. It has become the godfather to all local gun-selling advocates worldwide and has even opened a U.N. office to dampen that body's growing interest in taking guns out of circulation. So it's not enough to quash attempts to limit handgun violence in the United States, you see. The "happiness is a warm gun" lobbyists have fanned out across the globe, bringing their special brand of mayhem to all the world.

67   The Self-help Mania

Possibly because of its size, or variety in its publishing industry, or the personal emphasis of the news media, or the individualist ethic, America has given rise to a self-help craze that is now an industry of gargantuan proportions spanning the globe.

"Self-help à la Oprah and Dr. Phil is sweeping China," one news magazine reported in 2002. "A leading psychologist sees the development of an entire class of Chinese 'whose bellies are full, and now they're thinking about finding happiness.' From self-help books to courses in positive thinking, a new industry has sprung up to feed their hunger for self-exploration."

In Britain, says *The New Statesman*, "the feel-good industry is flourishing. Sales of self-help books and CDs that promise a more fulfilling life have never been higher," about $140 million in the United Kingdom in 2003, on per capita par with America's $600 million in purchases. "FranklinCovey, the public company formed by the author Stephen Covey, recorded sales of $333m in 2002. Spin-offs have included lecture tours, seminars and further *7 Habits* titles, including *The 7 Habits of Highly Effective Teens*, itself a bestseller, from Covey's son Sean."

It is an easy target, the self-help industry, with its one-minute managers and twelve steps or seven habits or forty-five days to a slimmer you. And it's come in for several rounds of ridicule and deconstruction, though it keeps sailing through all the storms. One critic, Steve Salerno, makes the basic point by quoting Archie Brodsky, a senior research associate for the Program in Psychiatry and the Law at Harvard Medical School. "'Psychotherapy has a chancy success rate even in a one-on-one setting over a period of years,' observes Brodsky, who coauthored (with Stanton Peele) *Love and Addiction*. 'How can you expect to break a lifetime of bad behavioral habits through a couple of banquet-hall seminars or by sitting down with some book?'" And Wendy Kaminer, one of our sharpest social critics, wrote more than a decade ago in *I'm Dysfunctional, You're Dysfunctional*, that people reading these books are willing to make major life changes based on "something their aunt or auto mechanic could have told them."

The books are followed by videotapes, CDs and DVDs, infomercials, speaking tours that can fetch six-figure fees, and even larger operations. One follows the next from the authors who apparently have a few more things left in reserve. Many of them are linked in some way to churches, and indeed the motivational thrust of self-help bears a striking resemblance to the inspirational. Many evangelical churches have been on a parallel track, replacing fire-and-brimstone preaching and soul saving with advice on how to better manage your marriage, children, finances, and the like—in effect, Jesus as the ultimate therapist.

The entire phenomenon is centered on the self and personal relationships, of course, and how to reform. The fundamental premise is that something in our childhood or a similarly vulnerable period has damaged our self-esteem, confidence, creativity, or entrepreneurial spirit. This can be repaired by self-discipline in one form or another, being kind to others, thinking ahead, sticking by some principles. Like Kaminer said, something your aunt might have told you.

The most troubling aspect of self-help is its wallowing in the personal. The "self" is all. The problems of life can be fixed by forthrightness, faith (in yourself, if not Jesus, too), and the resurrection of self-confidence. While some authors will nod toward the social context, the focus perforce remains on the individual—that's the genre, and that's the consumer. It is perfectly understandable that this "movement" grew up in America, with its long-standing emphases on individualism and the "pursuit of happiness." But the idea that virtually all personal problems are the sole responsibility of the individual, perhaps the nuclear family, is another large step away from the social and political communities that give our lives meaning and satisfaction, and enable us to solve problems together—some of them being the problems that self-help manuals claim to address, like loneliness, alienation, disempowerment, feelings of inadequacy, and lack of purpose.

This would seem obvious, but the therapy drumbeat never stops. And now, infecting China and England and everywhere else where books and CDs are sold, we can imagine the American obsessions with personal satisfaction inflaming unfulfilled souls worldwide. It fits rather neatly with all the other atomizing influences of American culture and economic policies, however coincidentally. But mostly it simply reveals Americans, some of us, anyway, as shallow and self-absorbed, caring little for our neighbors, much less the rest of the world.

68 Destroying the Left, Bankrupting Democracy

A favorite Cold War game of the right wing and nearly every president has been to undermine left-wing parties and other social movements around the world. The rationale stemmed from the purported link between such politics and the Soviet Union. In the paranoid view, they were fellow travelers of Soviet communism, if not in the pay of Moscow, or dangerously sympathetic to the communist agenda and therefore worthy of destruction. In the more mainstream version, leftists are seen as encumbrances; pro-business and pro-American alternatives are always preferred.

So for the entire post-1945 period, the United States has worked furiously through covert action, propaganda, foreign aid, and other instruments of foreign policy to destroy the left. This has included the democratic left, that which was neither communist nor authoritarian, but more akin to Scandinavian democratic socialism—and by far the largest segment of the left in any case. It is a policy that remains very much at work today.

The backwardness of this policy is obvious. Destroying any peaceful segment of another nation's political variety is a violation of their sovereignty. It robs them of political pluralism, a cornerstone of democracy. It deprives them of policy solutions, talented political leaders, social capital, networks of political activism—all the things that enrich a nation's politics and make it healthy and functional.

(There is a domestic corollary, "defunding the left," which is to say, attacking programs for the poor and middle classes that might strengthen ties of awareness and joint action.)

Of the top ten wealthiest countries, eight are governed by social democratic systems, and have been for many years.

So destroying social democrats is obviously bad for the targeted countries. Where did this happen? Anywhere political contests were under way, essentially, including Western Europe. The left was destroyed in Iran in the 1950s, 1960s, and 1970s, forcing dissent to be expressed by the mullahs then in the mosque and the souk, rather than the halls of power. Often, the attacks on leftist political parties drive some to guerrilla groups, as happened in nearly every Latin American autocracy backed by the United States. Assassinations of politicians like Patrice Lumumba in Congo and Orlando Letelier of Chile, the thousands in Indonesia in the mid-1960s, and others unknown to us have not only been immoral on the face of it, but in the long run deprived these countries of a viable political leadership representing a large—perhaps majoritarian—segment of the populace.

Today we see the same tendencies in foreign aid—where "democratization" efforts include building political parties along the lines of our own—and covert action, such as destabilizing Chávez in Venezuela. (By contrast, a foreign government or even a foreign citizen cannot legally support political parties here.) American corporations routinely try to influence politics in foreign countries where they have business, and they will, of course, always favor the conservative elites who will do their bidding. Global media corporations are particularly adept at this, casting aspersions on leftists the moment they come close to power. The leftward drift in elected governments in Latin America is now earning such opprobrium. It's an old habit, so common that we barely notice its relentless machinations.

The old democratic socialist parties are not the only enemies the U.S. government targets anymore, however. It must now take on the broad social movements that challenge marketization in particular—the so-called antiglobalization groups, transnational in character and highly effective in casting doubt on the agenda of the World Trade Organization, the IMF, and U.S. foreign policy. Whether or not these movements have been infiltrated, spied on, disrupted, or otherwise subjected to the usual bag of dirty tricks is difficult to say—it is highly probable—but the corporate press and think tank culture has pounded them without mercy ever since they successfully raised questions at Seattle and other venues of protest. Marginalizing dissent is another hoary trope of the right.

The attempt to destroy the left—it continues to thrive in those successful countries and elsewhere—is perhaps the best evidence of the narrowness, not to say disingenuousness, of the mainstream discourse on democratization. You see, we promote democracy only if it conforms to certain narrow definitions—pro-American, pro-market. If you have other ideas, watch out.

69   A Most Christian Nation
What Would Jesus Say?

America, the most religious country apart from India, and the most devout of Christian countries, has some, uh, contradictory attitudes when it comes to the teachings of The Man. The rest of the world is noticing. But maybe there is an explanation.

George W. Bush has said that he considers Jesus to be his most trusted political adviser. You would think that during the presidential debates someone would have explored this. So let's take up that inquiry now.

What, Mr. Bush, do you make of the following utterance?

"How hard it will be for those who have riches to enter the kingdom of God! . . . It is easier for a camel to go through the eye of a needle than for a rich man to enter the kingdom of God."

Mr. Bush, a rich man and the son and grandson of rich men and the very very good friend of many many very rich men, and indeed such a good friend as to enrich them further with tax cuts and contracts and other government goodies, would seem to be in hot water with this one. There's no escape clause like "unless they are the righteous," or "except for those who create new wealth that will, lo and behold, trickle down to succor the poor." No, it's pretty straightforward. "You cannot serve God and Mammon."

Let's try another. "If anyone strikes you on the right cheek, turn to him the other also."

Did Jesus forget this one when counseling you about bombing Afghanistan or offing 100,000 civilians in Iraq? Or did He confide, "I said that before 9/11."

Let's move on. The first words of Jesus quoted in the Good Book are from the Sermon on the Mount, a beautiful place in Galilee, and one can see why He might get carried away. But still. Here are some of the more memorable tidbits:

"Blessed are the meek, for they shall inherit the earth." Is this why the tax cuts all go to the rich, so they can prepare for this sudden turnaround? Clever you.

"Blessed are the peacemakers, for they are the children of God." This is a theme. He comes back to it later, and it's hard to miss. "He who taketh up the sword shall perish by the sword. Think ye that evil can be overcome by evil or violence by violence? The way of peace requires courage and patience, but it will prevail."

Or, even more to the point: "Mortals go to war so that they can inherit dust. It is because their vision is distorted by the followers of the lie that they value that which is nothing. In destruction there is no victory but for darkness. The power of victory is not force but Love." Did He say, there is no greater love than what comes out of precision-guided ordnance raining down on a village of terrorists?

Faith-based initiatives? National prayer breakfasts? Prayer in schools? "And when thou prayest, thou shalt not be as the hypocrites *are*: for they love to pray standing in the synagogues and in the corners of the streets, that they may be seen of men. Verily I say unto you, they have their reward. But thou, when thou prayest, enter into thy closet, and when thou hast shut thy door, pray to thy Father which is in secret." But what about the base?

There has long been a problem with American Christians of the evangelical kind—they believe what is convenient to materially acquisitive, socially restrictive, and globally belligerent political agendas. "Dalmatian Christians," one religious thinker calls them—they're Christians just in spots. The profound lack of self-awareness and truthfulness in this "faith" is a permanent canker on the national soul, visible for all the world to see. And, as is often reported, the president and his kind—there are now dozens in Congress and many more aiming for high political office—see their global role in apocalyptic terms, a message from God. One would like to think this is ludicrous, but it has too many serious consequences to laugh it off.

But then, "Beware of the scholars. . . ." There are a few things that must truly comfort George W. Bush in the Scriptures.

"The children of this generation revere the dead prophets while rejecting the living. So hath it been in every generation. The children of those who persecute you will build monuments to your memory."

Now that's more like it.

70   The Deadly Reach of Patio Furniture

Teakwood is *the* choice for elegant outdoor dining. It resists pests, cracking, and rotting, and ages beautifully. Teakwood, used in shipbuilding by the ancients of China, should last for decades with minimal customer care. Americans buy it by the boatful, and then some.

Trouble is, a lot of the teak in patio furniture is timbered in Indonesia by clear-cutting, much of it illegal, where ecosystems are devastated, habitats ruined, biodiversity threatened, and local economies and ways of life undone by unscrupulous corporations. Says one authoritative report: "The U.S. alone imported over $450 million worth of timber from Indonesia in 2002. Based on an illegal logging rate of 70 percent in Indonesia, the inference is that the U.S. imported over $330 million worth of timber stolen at source in Indonesia in a single year."

Even worse, perhaps, is the prospect that the teak came from Burma, logged through forced labor, exported to Thailand or India, milled and sold to American manufacturers as having come from a legal source. And teak, no matter how it is harvested, is considered to be an endangered resource.

The example of patio furniture—our own Crate & Barrel dining set is "plantation grown," not wholly reassuring—demonstrates, alas, how all things are connected. A simple choice of furniture is fraught with the possibility of having taken a precious resource from a third world country at prices no doubt exploitative, with labor that is too cheap to contemplate, leaving ruined hillsides, slag from badly run mills, and orangutans and hundreds of other species without their natural environs. Then there are the macro issues: deforestation adding to the woes of climate change, too few old forests to protect the world's reservoir of biodiversity, soil erosion, and goodness knows what else.

"Illegal logging is estimated to represent 73 percent of log production in Indonesia, 80 percent in Brazil and 50 percent in Cameroon," notes an investigative report. "It is clear from the illegal-logging rates in tropical timber exporting nations that a vast quantity of black market timber is being traded around the world, representing at least half the global trade." This activity targets the most ecologically sensitive areas and is "high impact" logging that does nothing to restore the habitat, soil, or anything else resembling nature.

Different kinds of resource exploitation go hand in hand. Mining and oil drilling often follow, or are followed by timbering. Once roads are built and ports fitted and such, the land and resources are easier to get at. Clearing for agriculture goes along with logging as well, even if, as in much of the Amazon, the soil cannot sustain crop growing or grazing for long.

And it's not just in the faraway lands that these outrages are occurring, but throughout North America and Central America, too, mainly logging and associated development.

But Asia remains the main source of many desirable woods, and American consumers are increasingly demanding the wood Asians export. "U.S. furniture imports from China alone have risen from $21.5 million in 1989 to nearly $1 billion in 1999, according to federal trade figures," explains a report in the *Philadelphia Inquirer*. "The United States is also a top importer of some raw tropical woods. About 60 percent of the world's mahogany is shipped to America."

This is a case where the culprits are not even predominantly American, and even the major consumers as a whole put America in a minority. But the growth of the American market is striking and puts tremendous pressure on even the most ecologically minded importers. And the Bush administration has done little to provide leadership. Its "Healthy Forests Initiative," a response to slightly higher-than-normal forest fires in

2002, was an invitation to logging companies to go into public lands to clear away the "fuel." Subsidies for logging, including some rainforest and old-growth forests, remain high. Tropical forests are disappearing worldwide at an annual rate of 50,000 square miles, so more action is needed, and it's not coming from Washington.

It is a case, too, of consumers' mindlessness. In an acquisitive culture, it's difficult to ask the hard questions about where something is made or how it might be less than a good idea to use it in a product. And that's how patio furniture becomes a planetary menace.

71   Public Diplomacy
Just Put on a Happy Face

Like a mantra after 9/11, earnest public figures of all political stripes began to intone a "solution" to Why They Hate Us: *Let us tell you how wonderful we are.* They don't know the real America. We haven't been getting our message across. If they only knew us, they would really like us and not fly planes into our skyscrapers.

From Richard Holbrooke, the Democrats' would-be secretary of state, to Karen Hughes, President Bush's cheerful emissary to the Arab world, the message is repeated over and over: Osama bin Laden and his ilk are winning the war for hearts and minds because they ("they" being Muslims, all of them apparently prone to support terrorism if we don't educate them) are unable to appreciate our fine, durable values. We are a religious people! We are the beacon of liberty! The citadel of democracy! We are rich, and we deserve to be rich, because we are good and true.

"Call it public diplomacy, or public affairs, or psychological warfare, or—if you really want to be blunt—propaganda," Holbrooke opined in the October 21, 2001, op-ed page of the *Washington Post*. "But whatever it is called, defining what this war is really about in the minds of the 1 billion Muslims in the world will be of decisive and historic importance."

(The war, Dick, can also be defined as the U.S. interest in oil, dropping bombs on villages, locking up men and throwing away the key, and torturing quite a number of them, too. But I digress.)

Ms. Hughes, undersecretary of state for public diplomacy, appointed as the debacle in Iraq was becoming clear to all, is a kind of debacle all by herself. ("I'm really, really enjoying my new job," she said after two months. "I spent the four months before that studying." *Four whole months*.)

In her first major foray abroad, Undersecretary of State Hughes visited Egypt, Saudi Arabia, and Turkey. It went so badly that even the sympathetic *Weekly Standard* lampooned her as "Karen of Arabia," noting that "Hughes was only two days into her five-day 'listening tour' of the Middle East, and she was relentlessly 'on message.' Her unshakable discipline in sticking to the script has a mind-numbing effect when you watch her through several events a day.

"'I go as an official of the U.S. government, but I'm also a mom, a working mom,' she told reporters on the flight from Washington to Cairo." She emphasized America's religiosity again and again: "It's in our constitution, 'One nation, under God.'" She met with selected audiences, much as her boss rarely veers away from military bases and Republican fund raisers in his public appearances.

"If you set out to help bin Laden," said terrorism expert Robert Pape, "you could not have done it better than Hughes."

"Put the shoe on the other foot," writes Fred Kaplan in *Slate*. "Let's say some Muslim leader wanted to improve Americans' image of Islam. It's doubtful that he would send as his emissary a woman in a black chador who had spent no time in the United States, possessed no knowledge of our history or movies or pop music, and spoke no English beyond a heavily accented 'Good morning.' Yet this would be the clueless counterpart to Karen Hughes."

But let's play fair with Hughes. She is not only following instructions, she is playing with a deck that does not have any-where near fifty-two cards. Try to convince the Muslim world that we're okay in the midst of two U.S. wars in Muslim countries, a long history of support for Israel, years of Hollywood stereotyping of Muslims (continuing, unabated, as I write this with *24, Looking for Comedy in the Muslim World,* and others—talk about clueless), and an official antiterrorism campaign in the United States that is recklessly and needlessly targeting Muslims.

The real problem, perhaps, is not that they don't know us, it's that they know us only too well. American television, mov-ies, books, magazines, corporations, products, advertising, tourists, newspapers, and missionaries are everywhere, and have been for years. We dominate the global news and entertainment media. We have been pouring government-sponsored information into Muslim societies. We have been absorbing or rotating Muslim immigrants or temporary workers or students in and out of our society by the millions.

The terrorists are either the relatively sophisticated men who have lived in the West (like many of the 9/11 attackers), or are enraged by U.S. actions in the region. Karen Hughes telling them that she's "a mom who loves children" is not going to alter either of those phenomena.

But there is another, fundamental flaw in the public diplomacy idea. The real problem does not reside in the rest of the world because it does not understand Americans (and therefore we must tell them how fine a people we really are). The problem is that we are constantly telling them this, but we don't listen in return. We know so little of the developing world in particular that we could not possibly grasp that hatred could mount to such a point that a 9/11 attack could not only happen, but that it would be treated with outright glee or a nod of "they finally get theirs" in many quarters of the global south. And that hatred, or disgust or disappointment, is based not on misunderstandings, necessarily, but on the sometimes accurate perception of an America that cares only about itself, enriches the wealthy at the expense of the world's poor, and belittles their aspirations, their cultural preferences and religions, and their politics.

We know, they don't. We have lessons to teach, not to learn. We have values to convey, not to rethink. It is an endless one-way street.

Public diplomacy, in short, is a substitute policy, a red herring. It's one of those easy fixes (remarkably hard to do, it seems) that is wholly self-gratifying—let's tell them about Martin Luther King!—and allows policy makers to avoid the hard choices about global equity, American bullying, and corporate rapacity.

Forget public diplomacy and act fairly; the hearts and minds will follow.

72   The Killing Fields of Death Row

Few more barbaric acts remain in the quiver of governments than execution. Killing someone who may have committed a crime is the ultimate abuse of power. It is a deliberate act of vengeance and no more. Among civilized nations, the United States is alone in persisting with this practice, and it tarnishes our standing as a leading light of human rights.

The arguments are well known and the facts are easy to

find, so I won't belabor the obvious. Capital punishment does not reduce or deter violent crime. Many innocent people have been executed (120 condemned people were released from death row since 1973 after their innocence was proven). The racial discrimination in who gets fried or gassed is appalling. Fifty-nine people were killed in 2004, and 3,400 remained on death row in late 2005.

It's an easy case to make—the immorality and ineffectiveness of the death penalty—but the United States, or, more precisely, the thirty-eight states and the feds that retain the killing fields, have put America in some exalted company globally. Here are the top ten executors in 2004 world wide:

China 3,400
Iran 159
Vietnam 64
United States 59
Saudi Arabia 33
Pakistan 15
Kuwait 9
Bangladesh 7
Egypt 6
Singapore 6
Yemen 6

Now, aren't those the kind of countries you want to be associated with in the human rights derby? Fully 122 countries have abolished or do not practice the death penalty.

It is not just the embarrassment or injustice that should concern sensible people, however. A more concrete cost results from how other countries view our hand on Old Sparky. "One area where the U.S. has frequently sought cooperation is the exchange of prisoners," notes Richard C. Dieter of the Death Penalty

Information Center. "The U.S. has extradition treaties with many countries whereby the country which detains a defendant charged with a serious crime in another country will return that defendant to the charging country. An exception, however, is made in capital cases. Countries such as England, France, Canada, Mexico, Italy, the Dominican Republic, and Germany have refused or delayed the extradition of people charged with murder in the U.S. in order to secure assurance from the prosecution that the death penalty will not be sought."

There are four international treaties, numerous resolutions passed by international organizations in which we are members, and countless pleas from our friends around the world to stop this ridiculous and shameful practice. Gradually, law and opinion is changing in America toward common sense, but few political leaders see opportunity in taking up this issue.

As long as we resist the worldwide opprobrium on executions, it's much easier for China, Iran, and other harsh regimes to get away with it too. If the United States did the right thing and placed a moratorium on executions, or even outlawed them, it would bolster our moral standing and improve the chances of ending this Dark Ages mentality in other places.

Our neighbor Canada (where the murder rate dropped dramatically after the death penalty was abolished) has some wise words on this matter. "My primary concern here is not compassion for the murderer," Pierre Trudeau, the former prime minister, once explained. "My concern is for the society which adopts vengeance as an acceptable motive for its collective behavior. If we make that choice, we will snuff out some of that boundless hope and confidence in ourselves and other people, which has marked our maturing as a free people."

73   The New Age

That would be the Age of Aquarius, which in fact may not begin until 2060, or 2600, despite the Broadway hit some thirty-plus years ago. Or it already began, probably in 1968 or 1997. Or perhaps it was, well, it's a little indistinct. But it is a New Age.

And you can tell that by walking into any of the 5,000 New Age bookstores in the United States. Five thousand. And picking up any of the hundreds of magazines (now being outdistanced by the Web, where Age of Aquarius has more than 2 million hits and New Age magazines has 35 million). It is hard to avoid.

It is also hard to describe what New Age is. Astrology seems to be big. Alternative stuff—solar energy, homeopathic medicine, herbal things, living arrangements, living quarters, spirituality, foods and food production, hair, relationships . . . the list is comprehensive. It speaks to a longing for something different from mainstream culture, economics, and politics. Altogether, it's pretty harmless. But its vastness is impressive, and anything so phenomenal needs to be looked at a little skeptically.

Now, I bring sympathetic skepticism to this task. I have close friends who are involved in some of these activities and beliefs. As a child of the sixties, I acknowledge that I sometimes flirted with some of the practices I am about to ridicule. But that was long ago, when the Age of Aquarius and I were both younger, and as we know, youth is wasted on the young.

New Age subcultures have enlivened and sustained many things that are quite useful and enjoyable—traditional forms of music, for example, and a broader appreciation of old traditions. Practices like meditation are a godsend to many. Emphases on organic farming and food have obvious merits, as does the general embrace of outdoorsiness. Holistic thinking also has its place.

So many of these influences have been mainstreamed that they are now difficult to recognize as being promoted, significantly, by what we would call counterculture or New Age people.

But there is a side to New Age that is just plain loony. The astrology and all that goes with it, the hybrid spirituality and obsession with the "other world," the phony herbal medications (not all, but most) and much other healing nonsense, the idealized comic-book mythology, the self-help mania, and, more troubling still, the intentionally fringy politics. The latter comes, often, with a conspiracy theory mentality that explains everything and nothing at the same time. The irrationality of it is breathtaking. It's where Woodstock and the Dittoheads meet.

The pity of it is the apparent need to believe in things that this world does not have to offer. UFOs and a fixation with extraterrestrial life, the magic of crystals and communing with the dearly departed and such. It's global in scope, and while a lot of the traditions and inspirations come from overseas—from the Maharishi Mahesh Yogi (yes, my master circa 1970) to Gurdjieff to Lao-tzu to tae kwan do—it is as always America that absorbs, reframes, and repackages it—hodgepodginization, one might say—for mass consumption.

Isaiah Berlin, the English philosopher, spent much of his intellectual effort on explaining the origins and evolution of Romanticism in Europe, and its ultimate influence on fascism. It was the essential irrationality of it, the belief in grand myths and the suspicion of the Enlightenment, indeed, the seizure of a superior attitude toward rationalism and empiricism that grew in a soil of victimhood and marginalization, which made it so powerfully alluring and dangerous. Kind of like evangelicalism today.

The New Agers, far more benign than these other cultures of unrational mythmaking, are nonetheless wallowing in shallower pools of the same waters. That's a little bit sad, mostly harmless, but not something to be celebrated as America's regifting to the world.

74   Committees of the Imperious

We now have decades of experience with a well-worn device of elites of all stripes trying to couch their opinions as something more authoritative. That's the committee or commission or project or foundation or whatever on the present danger or the new American century or defense of democracies or intelligence summit or something else very, very dire.

Many of the committees have a common message. *America is in peril and we must assert our rightful role as the preeminent power of all time and get tough with Islamic terrorists, spineless Euros, and antiglobalization freaks. It is our obligation, our destiny.*

Corporate opinion and revolving-door government elites have long had their places to meet and forge consensus—the Council on Foreign Relations, the Trilateral Commission, and so on. Think tanks can serve a similar function. But these newer, flimsy committees of American imperialism—I wish just one would cop to that name—are mainly letterhead operations with a Web site, a "committee" of members, directors, councillors, and the like, and a publicity machine. The Project for a New American Century garnered an unwarranted amount of attention when it formed in 1997 by calling on President Clinton to knock off Saddam—what a great idea *that* was—and reestablish American primacy, as if it were somehow last. Several of its cohorts (one can imagine the candlelit dinners in McLean, with the prime rib and Merlot compliments of Lockheed Martin) ended up in the Bush administration, not least Mr. Imperious himself, Dick Cheney, and *presto!* We're off to Baghdad.

While there is little if any institutional substance to these places, they do have "staff" that generates op-eds to influence public opinion and boost allies in Congress. So we have the director of the Project for a New American Century, Ellen Bork (yes, *co-incidentally*, Robert Bork's daughter), publishing in the *Financial Times* and the *Asian Wall Street Journal*. Cofounders William Kristol and Robert Kagan are masters of this game. The Project has mainly been fixated on Iraq, where one might say that the New American Century met its end. (Among the three reports it has listed is one from April 2005, entitled, "Iraq: Setting the Record Straight." Read all about the al Qaeda–Saddam connection.)

The modern-day precursor of New American is the Committee on the Present Danger, which is now undergoing its second revival. Created in 1950 to pump up a nation not properly anticommunist enough, it made its biggest mark in the late 1970s when it warned incessantly of the imminent Soviet victory in the Cold War. It was made to upend détente and boost military spending. Like its progeny, the Project for a New American Century, it created a frame for debate and was even more successful when a pliant and inexperienced ideologue became president, ripe for capture.

Today, the Committee has as honorary cochairs Democratic senator Joseph Lieberman and Republican senator John Kyl, along with George Shultz, Reagan's second secretary of state, and the omnipresent James Woolsey, former director of the CIA, and a rogues' gallery of members, many of them appearing on the other committees and projects and summits and whatnot. It is dedicated to "winning the global war against terrorism and the movements and ideologies that drive it." There does not seem to be a staff, or research or reports. There is an e-mail mechanism to request interviews, and a P.O. box address. I suspect the Web site may actually be a parody stealthily placed by the writers of *The Daily Show*.

We might soon expect the Committee to Restore the Shah, Project for a Compliant World, Council on Invading All Muslim Countries and Converting Them to Christianity, and the New

American Foundation for the Elimination of All Remnants of Socialism.

This would be laughable if it were not for the fact that the committee or project or foundation for a new America or the present danger or whatever does exert influence through the news media and the dense web of political networks these mechanisms have accumulated over the years. They appear overseas as the graybeards of American private diplomacy. They junket and op-ed and dine and lobby and opine on Sunday mornings and beat the drum incessantly for more wars and the spread of American "values" and furrow their brows about weakness in Europe and the Security Council.

In effect, they have created a universe of talking heads as an alternative to those fuzzy headed know-it-alls at Harvard and Yale and Berkeley. Some of the well greased create journals and name chairs to complete the Potemkin village effect. Their essential relationship to the military-industrial complex is rarely acknowledged.

Opinion formation is the goal perhaps above all else; opinions within the right-wing elite, a kind of clubby consensus building exercise, and then in the broader news media. It's clever tactics. A pity they have been wrong about every major challenge—overestimating Soviet strength, coddling the wrong dictators, ignoring genocides, asleep at the wheel before 9/11, starting high-casualty wars. But the prime rib and Merlot are always perfect.

75   Forgetting History

When Ronald Reagan went to Bitberg, Germany, in 1985 and gave a speech at a graveyard filled with Nazi officers who carried out the Holocaust, many Americans and Europeans were appalled by this apparent act of homage to the most vicious regime in history.

George Will, the right-wing television commentator, explained his hero's gaffe this way: I've been talking to German youth, Will told his ABC News listeners, and they are saying to me, "Forget history! Let's move on."

For an alleged conservative to say that one should forget history—very recent, very relevant history—is telling. Take one for the Gipper. But there is a tendency in American life that resonates with this. America is the "new land" in all meanings of "new," and among those is the capacity to leave the old behind. In some ways, this is refreshing and enlivening. I have seen, in the eastern Mediterranean and Bosnia and other places, how gripping and even paralyzing "history"—a memory of wrongs perpetrated by an adversary—often is, generation after generation. Many people have come to America to start a new life precisely to leave their own personal, family, class, ethnic, or national history behind them, because it is crippling.

But this tendency extends well beyond immigrants, into our political discourse. Forgetting history is a national pastime of the opinion and policy elite.

Consider Iraq. All the editorializing about the evil of Saddam Hussein in the run-up to the 2003 war rarely mentioned how Reagan saved him, Bush the first built his regional policy around him, the CIA supported him, and so on—it was as if he emerged from the head of Zeus in August 1990 to occupy Kuwait: that's when history began. John McCain's comment about not owing anything to Haiti pivots on the same kind of ignorance or neglect. When Peter Jennings was interviewing Iranian president Rafsanjani in the early 1990s, he began by saying, "At long last, Mr. President, don't you think you owe an apology to the American people for the hostage taking in 1979?" Rafsanjani ruefully pointed to the CIA coup against Mossadegh and a quarter century of support for the bloodthirsty shah.

There are so many instances of this forgetting that it begins to look intentional. American-led genocide in the Philippines in the early 1900s? Never heard of it. Ethnic cleansing of indigenous peoples in what is now America? We call that Manifest Destiny. Multiple military occupations of Nicaragua, Mexico, Cuba, Dominican Republic . . . ? Didn't get that in my schoolbooks. A nation of immigrants? We're shutting down the borders.

Political leaders will rarely acknowledge historical facts that are inconvenient to their policies or worldview. But others in the policy world and the news industry could—but they have agendas, too.

The pivotal reason for forgetting, apart from just not taking the time to sort out the complexities of history (which are not always that complicated, by the way), is that it requires setting a context that muddies responsibility. That is, Saddam can be blamed for everything if you start with his occupation of Kuwait or his gassing of the Kurds. Castro can look particularly nasty if you forget the decades of American support for corrupt dictators before him. We don't want contexts that intrude upon our conventional wisdom about things, especially if American culpability might be introduced as well.

Can't there be a balance between the paralyzing memories and politically willful forgetting? Discussing history, reminding people that there are contexts and reasons for contemporary events, would be a start. That might get in the way of the "reality shows" or infotainment or obsessions about the latest trend, but with a little work, history could be interesting as well as enlightening.

76   We're Number One

Among the more obnoxious tendencies of redneck culture in this country is to declare the United States as number one, without regard for what that actually might mean. It seems to arise with blaring bravado when the Olympics are held on our soil—chants of "U.S.A.!" abound—but it has a longer and broader pedigree than that. Is it so that every imperial power regards itself in this way, loudly and insistently, as if others who do not share the sentiment or citizenship might be persuaded by these giddy outbursts?

In total industrial output, in political clout, and in military spending there is no doubt that America is number one. But the way we care for our own might be a more telling set of criteria than the raw, aggregate numbers of power. So, consider a few:

- The United States is forty-ninth in the world in literacy, and is twenty-eighth out of 40 countries in mathematical literacy
- Our provision and *quality* of health care and medicine ranks near the bottom of industrialized countries
- Childhood poverty is more prevalent here than in all developed countries except Mexico
- The best run companies and most successful banks are predominantly European
- The proportion of scientific articles published by Americans is declining sharply
- Infant mortality—the United States ranks forty-first in the world

"In most important categories we're not even in the Top 10 anymore," writes Michael Ventura, who compiled many of these and other numbers, in the *Austin Chronicle*. "Not even close. The U.S.A. is 'No. 1' in nothing but weaponry, consumer spending, debt, and delusion."

As Ventura correctly observes, any politician who suggested we were anything other than number one would be committing political suicide. That these numbers have persisted for many years and gotten worse in most respects should be an occasion for some national introspection. But the opposite is the case.

This became especially acute after 9/11, when the reassertion of American pride and exceptionalism seemed necessary. Even now, one can find an enormous number of flag decals on cars (actually, mainly on SUVs and other light trucks) alongside the yellow ribbons. Many of them are emblazoned with the number-one proclamation. Doubtlessly, there is plenty of cheap psychology to explain this. Americans feel victimized (by what exactly is hard to say, but watch Fox News for ten minutes and you'll feel the self-pity) at the same time as they are exerting more bullying power than ever.

The Pew Global Attitudes Project, the best measure we have of how others in the world regard us, consistently demonstrates rather poor rankings for U.S. policies particularly, with the American people doing better on qualities like honesty and being hardworking. But most of the world regards America as greedy, violent, rude, and immoral. That's a rather full range of dislike. Attitudes about U.S. policies are even more negative. The most striking is the universal belief that America cares only about itself.

Anything wrong with this? Isn't being a global leader—world's only superpower, etc.—kind of like being the very best high-school football team in Texas, and we all want to yell, "We're number one"? Let me suggest a couple of alternatives.

The droning incantation of superiority is a distraction. If you think you're number one, all is right on God's green earth—and His favorite nation, of course—and not much needs fixing. Those childhood poverty numbers? Declining science prowess? Poor health care system? Hey, we're number one.

But it is also an affront to other people in the world, and not just in terms of rudeness. American leaders constantly insist on doing things our way—economic policy, particularly—because somehow we know what's best in all things. That this is patently untrue does not seem to matter, and it is this arrogance that creates such fissures among the industrialized democracies, not to mention the beleaguered global south.

"To make us love our country," the conservative English philosopher Edmund Burke observed, "our country ought to be lovely." America's aspirations and self-image are not matched by its actual achievements or intentions. It lacks, in Burke's sense, this loveliness. And because it could be so, the tragedy is all the greater that it is not.

77   Strong States, Weak States
Whose Side Are You On?

The average American pays little attention to those many countries that are neither in crisis nor close to the United States geographically or ethnically. It's not easy figuring out where or what a Kyrgyzstan or Côte d'Ivoire is much less how they fit into global politics. These places don't quite rise to the sad prominence of Rwanda or Iraq or Guatemala or Vietnam, where genocides were ignored, wars started, or dictators installed. This relative anonymity permits quite a bit of leeway for U.S. elites to create relationships with these states that work to the advantage of economic and political interests, though not always for the good of the people in that region.

All kinds of countries fall into this category. A weak state like Bolivia earns quite a bit of interest from the U.S. government because of its cocaine production. It has a "weak state"—its governing structure of executive, legislature, courts, police and military, and so on have been racked by instability, corruption, a variety of strongmen in the presidency, and so on. Most recently it has been locked in an internal battle for control between the elites who want to privatize the economy and sell off assets to foreign corporations ("structural adjustment," begun in 1985) and the two-thirds of the population, mostly rural, that live in poverty. Many of them were traditional coca farmers and resist eradication efforts.

As long as Bolivia pumps its natural gas reserves and wipes out the *cocaleros,* it's our friend; if it becomes populist—as it seemed to in late 2005 when it elected a leftist, Evo Morales—then it is a threat; in the words of a *Washington Post* columnist, "Once in office, U.S. officials fear, Morales could become more radical than [Venezuela's Hugo] Chávez and move a country that has deep political, regional and ethnic fissures to the brink of civil war." Of course, it is the populist who moves the country "to the brink of civil war," not the white elite that sold all the assets and polarized the nation in the first place.

Bolivia is fairly typical of the weak state quandary—those that can be manipulated because of the instability wrought by war (sometimes a vestige of the Cold War), economic weakness (structural adjustment, often), a nearby threat (war on terrorism), or just bad luck. They don't quite dip into the "roadkill" category—to recall the term of choice of a certain national security adviser—because they have some value, typically a natural resource to exploit, or some flat land for a U.S. Air Force base. A lot of Latin America falls into this category, as do a few states in Africa, much of the former Soviet Union, and a sprinkling in Asia.

The flip side of this is the "strong state" that also, remarkably, escapes much news media attention but is of interest to U.S. rulers. Pakistan was one such place before 9/11 put it on the map of American concern. Indonesia, the world's largest Muslim country, is another such place. Both had convenient relationships with Washington for years, where we would overlook their considerable human rights abuses in exchange for playing them against another bad guy in the region.

Perhaps the perfect example is Turkey. Created from the ruins of the First World War, when the doddering Ottoman Empire aligned itself with Germany, Turkey's heroic founder, Kemal Atatürk, built a nationalistic and centralized state. He modernized Turkey in many ways, but also created a mild form of fascism, which excluded anyone who was not a Turk. This led to numerous uprisings, mainly from its enormous Kurdish population, and indeed a civil war of greater or lesser intensity, in which 40,000 have been killed, resulted from those policies. Along the way, the Turkish military has held its exalted place in politics and society; the United States insisted that Turkey be admitted to NATO as a bulwark against the Soviets, and the military has never relinquished its political control. Formally it has seized power three times since 1960, and has intimidated every civilian government.

The U.S. government, and especially the Pentagon, has embraced Turkey as a valued ally and provided it lavishly with military hardware—it has the largest fleet of F-16s outside the U.S. Air Force—and political support. The civil war was fought with F-16s, Black Hawk helicopters, Abrams tanks, Bradley Fighting Vehicles, and all sorts of other American-made weapons. In the political wars over the division of Cyprus and Turkey's on-again, off-again courtship with the European Union, Turkey has had stalwart American backing.

Remarkably, its human rights violations have gone virtually without notice. During the civil war that was most intensive in the 1990s, one million Kurdish villagers were forcibly evacuated (to remove potential sources of support for Kurdish guerrillas), with nowhere to go. Can you imagine one million Palestinians or Irish or Mexicans being forced at gunpoint (American-made gunpoints) from their homes and not receiving any notice? They moved to shantytowns and flooded into displacement camps. Most of them remain there.

Since the nineties, Turkey has been liberalizing, it is said, but this would be lost on the dozens of writers and publishers who are being prosecuted for speech crimes. Violations of Kurdish rights continue. Denial of the Armenian genocide is state policy. All of these insults to democratic process and ideals are violations of treaties Turkey has signed and ratified, including those with the United Nations, the Council of Europe, the Organization for Security and Cooperation in Europe, and NATO.

These multiple, repeated, violent, unapologetic assaults on people and law worry not our Washington elites. The money flows, trade escalates, enormous political pressure is exerted on European allies, etc. Turkey's our friend, period.

Or is it? The irony of strong state politics is that sometimes they turn on their patrons, and sometimes they suddenly devolve into weak or crippled states (think Congo, Nigeria, Argentina). Pakistan, so often favored over India in U.S. foreign policy (and one of the reasons India pursued nuclear weapons), has harbored al Qaeda and other jihadists, supports violent groups in Kashmir, and has not always been helpful with other U.S. adversaries like China. Turkey refused to support our forays into Iraq, has friendly relations with Iran, supported jihadists in Chechnya and roils the stability of the Caucasus, and refuses to deal on Cyprus. This is the return-on-investment of billions in giveaways and nodding at massive human rights violations, military coups, and dissing democracy.

Like so many aspects of U.S. foreign policy, the mindless support for narrowly conceived American interests in the anonymous weak and strong states of the world backfires time and again. It is sometimes said that foreign aid is the transfer of wealth from the lower middle class of rich countries to the upper classes of poor countries. From the strong to the (apparently) strong, it is usually a military-to-military lovefest.

78   Fat Country, Fat World
The Magic of Processed Food

It began with a Frenchman, wouldn't you know, and the needs of an imperial army. Napoléon wanted a way to preserve food for his soldiers on their long treks across Europe, and he offered no less than 12,000 francs for the new invention. Nicholas Appert, a brewer, confectioner, and chef in Paris, came up with the idea of canning food, and perfected it over fifteen years. It was, in essence, the first mass-producible, transportable, processed food technology.

And what an invention it was. For people living away from the major agricultural centers, or merely wanting to preserve food and have some variety, canning and its many progeny were a godsend. It continues to be an important way to feed people reliably. But processed foods have moved well beyond simple canning and other forms of preservation.

Today, the processed food industry is enormous and has global reach. Processed foods are widely considered to be the root of America's growing obesity problem, with high-salt, transfats, excessive and unhealthy sugars (such as high-fructose corn syrup), and too little of what is known to be healthy. And now we're marketing and making these products worldwide.

"In spite of largely saturated markets in all types of processed foods and beverages in recent years," writes a Tufts University nutritionist, "the food industry as a whole continues to grow both in sales and product volumes. This economic paradox of continued growth in spite of apparent market saturation results in the caloric source of much of America's pandemic obesity." America's consumer food market, valued at $920 billion in 2003, has expanded the per capita calorie intake by 25 percent since 1970. So the food corporations' bottom line grows along with our waistlines.

The growth of the processed food industry here and in Europe paralleled growing personal incomes and the demand for easier cooking options at home and more dining out of the home. Several inventors and countries around the world have contributed to the phenomenon of unhealthy processed foods, but as is usual in these situations, Americans have perfected the product.

Something like 40 percent of the world's top food manufacturers are American, and our exports equal about $27 billion, one third of which are meats. But the more potent influence of American business is direct foreign investment rather than exports. Major firms like Kraft and Wal-Mart are either producing processed foods abroad or marketing them in the growing numbers of retail superstores sprouting up in developing countries.

"In 2002, Heinz expanded its plant capacity by 15 percent in China and opened a new plant in the Philippines," notes one study. "The Kellogg Company now has manufacturing plants in China, India, Japan, South Korea, and Thailand for supplying retail chains in Asia. PepsiCo, the second-largest U.S.-based food company, is continuously extending its geographical reach with its extensive international marketing arm in snack foods, currently focusing on Latin America and Asia-Pacific." Global processed food sales are now $3.2 trillion, three-quarters of all food sold.

One might say, it's about time the rest of the world got fat too. And why not? It will be a boon to the pharmaceutical companies. Really fat Chinese people may not be interested in devouring Taiwan. A Twinkies-rich diet for the Middle East may reduce terrorism. Who knows?

But if one is a little bit concerned about health and nutrition, the explosion of processed foods in the developing world is generally not a good thing. It's one of those problems in which no one institution or person can be blamed. It happened in part because we rely on the giant private companies to provide us with most of our foods, and they naturally gravitate toward what is cheap to produce and tasty. And, sadly, these nasty things do taste good. The U.S. government has subsidized many of the worst offenders—production of corn, as noted elsewhere—and its nutritional efforts are pathetic. School administrators, desperate for revenues, put vending machines of soft drinks in the hallways for kids to develop deplorable habits.

If we can't save ourselves, how can we be expected to bring the good news of healthy foods to the rest of the world?

79   The Filthy Rich

This is an old saw, isn't it? But the rich just keep getting richer, and America remains the land of the very wealthy. Half of the people in the world with net assets in excess of $8 billion or so are Americans. About half the remainder are dependent on the United States (like oil sheikhs), and most everyone has assets here.

The very rich lead sumptuous lifestyles that have come to symbolize the American Dream. *Sumptuous* does not quite capture the essence of the over-the-top quality of spending. Conspicuous consumption has always been a hallmark of the very wealthy, but now it is global in scope. Houses are owned throughout the chic world. Banking, finance, property holdings, and other business enterprises know no borders. Culture, values, and outlook are shared by people of like income, not by nationality or religion.

Hollywood does its best to humanize the wealthy and convey

the repeated message that mountains of gold do not add up to happiness. There's always the Hugh Grant– or Richard Gere–played character, suave but hollow, who is fulfilled by the shopgirl, secretary, or escort. It's a pleasant fantasy for everyone.

All of this would be relatively harmless were it not for the determined self-aggrandizement of the superrich. While there are exceptions—George Soros, Bill Gates, Ted Turner, among others—the rich are not very generous. Much less generous, in fact, than the poor, who give a considerably larger percentage of their income to charity than do the very wealthy. And they are not only lacking in basic sentiments of doing a little good with all the good fortune that's come their way; they want *more* fortune.

Hence the pressure for tax cuts, cloistered banks and permissive financial rules, government subsidies and protection, and all sorts of other schemes. Of those top fifty fortunes with more than $8 billion each, we have the Waltons of Wal-Mart, we have oil, chemical, and pharmaceuticals, bankers and other financiers—CEOs in industries and companies that have fed at the public trough for quite some time, and keep going back for more. As I detail elsewhere, many of these industries have inimical impacts on the world.

The rich have pushed for the credo of individualism and less public diddling with the economy (unless they need more government diddling to aid their companies). Some even write books of self-praise, typically extolling free enterprise and the kind of gumption that got them their first billion. That this is out of reach for all but a handful of the lucky, and is an inappropriate philosophy for ordinary people, never seems to dawn on them. Perhaps it's a necessary brace for the rapacious business practices and manipulations of public policy through which their fortunes have in fact been built.

We should not forget that it is not only the trashy rich who shop and ski and party endlessly that we're talking about here; it's also the run-of-the-mill CEO, the top 100 of whom earn in the vicinity of $33 million annually. They, more than anyone, promote the ideology of wealth.

The get-rich-and-live-like-us mantra is especially pernicious in the United States, where real income for families has not gone up for thirty years, despite the apparent prosperity of the 1990s, stock market rallies, steadily improving worker productivity, and low inflation. Income disparity is growing, so the rich get the fruits of the economic growth and the middle classes and poor get two income earners and long hours at work. In Japan, no slouch when it comes to economic growth, executive pay is 11 times the average worker. In Britain, it's 22 times. In America, the CEO is likely to earn 475 times the average employee of the company.

As a result, the ordinary folks typically are not inclined to generosity—toward immigrants, foreign aid, or economic policy generally—when they feel a tightening belt. So the politics of international development and other global issues that require American support have become more and more miserly. Altruism sparks up when people feel flush, and the great mass of the American people have not had a string of good years since the Beatles ruled the waves.

So the wealthy stick it to the rest of the world in every available way. Bad example, poor charity, government favor, manipulating finances, hiding assets, pushing for tax cuts, and on and on. We've perfected this "filthy rich" thing in America, and they have taken the rest of us to the cleaners.

80   The Global Gag
Family Planning the Extremists' Way

We have before us in a small corner of policy—small in the attention it gets—yet another illustration of how tiny-minded extremists have captured U.S. foreign policy making to the detriment of much of the rest of the world. Funny how often this happens.

It began as so many of these things do with Ronald Reagan and his so-called Mexico City policy in 1984, where he announced that the U.S. government would not allow NGOs to use federal funds if they provided, promoted, or advised on abortions. On his second day in office, after the hiatus of the Clinton years, George W. Bush reinstated the policy, which forbids use of non-U.S. funds for these purposes as well.

Family planning advocates call this the "global gag" rule, and for good reason.

By preventing NGOs from even discussing abortion as an option for women in the developing world, it jeopardizes their lives in some cases and certainly reduces their options. "Our continuing research shows the gag rule is eroding family planning and reproductive health services in developing countries," explains one of the world's oldest and most respected organizations, Population Action. "There is no evidence that it has reduced the incidence of abortion globally. On the contrary, it impedes the very services that help women avoid unwanted pregnancy from the start."

The suppression of counseling advice and services generally impedes professional care for women who are pregnant, and, in poor countries, may have few places to turn to for good care. "Mexico City policy guidelines on counseling and referrals are ambiguous and unworkable in the countries where abortion is permitted under a wide range of circumstances and therefore put women's lives at risk," reports *Science* magazine. "According to the American College of Obstetricians and Gynecologists, these restrictions 'violate basic medical ethics by jeopardizing a health care provider's ability to recommend appropriate medical care.'"

But that's not all. Bush has also withheld U.S. contributions to the United Nations Fund for Population Activities or UNFPA, which provides the bulk of family planning work in the world, because of allegations that UNFPA supported "forced steriliza-tions" in China. "What I find so outrageous is that Bush withheld this $34 million based solely on testimony from the Population Research Institute, an arm of a far-right group," Representative Carolyn Maloney (D-NY) told a *Salon* reporter. "PRI is the only organization that has ever made these allegations. The adminis-tration is going against the will of Congress and the international community by allowing a small band of extremists to hamstring its foreign policy."

As it turns out, however, the conjectures about China were refuted by the White House's own fact-finding mission in 2002. But the funding cutoff was not changed. And, as Population Action explains, "There are important reasons why UNFPA should work in China. Roughly 20 percent of the world's women live in China. UNFPA is a source of contact with international experts who can recommend alternative and voluntary approaches to China's com-pulsory family planning program. In early 1998, UNFPA's Execu-tive Board approved a $20 million, four-year program that aimed to promote voluntary use of family planning and to allow couples to freely choose the size of their families. The emphasis on volun-tarism represented an important breakthrough."

So even by its own standards—wanting to promote voluntary, family-based decisions making about family size—Bush should support UNFPA. In China alone, according to the authoritative Guttmacher Institute, the "'unmet [contraception] need'—deduced from the number of women using 'traditional methods' or no contraception at all—translates to 52 million unwanted pregnancies each year, bringing on 1.5 million maternal deaths and more than 500,000 motherless children." Read those numbers again.

Think China needs some help?

News item:

> *In September 2005, the Bush administration announced that, for the fourth consecutive year, it will withhold the United States' $34 million contribution to the United Nations Population Fund. To date, the Administration has withheld $127 million in funds already appropriated by Congress. One year's withheld funding of $34 million could prevent as many as 2 million unwanted pregnancies and 4,700 maternal deaths in developing countries.*

Is this what they mean by "compassionate conservative"?

81   The Imperial City

Has there ever been a more provincial capital of an empire than Washington, D.C.? Rome, Istanbul, Paris, London, even Moscow—cosmopolitan, outward looking, filled with artists and artisans, bursting with creative energy and industriousness, inventors and poets cheek by jowl, denizens of the outer reaches of the imperium flowing in and out like a great river. They had it all. It made for rich experience, a cross section of life itself and its higher aspirations.

Washington? Built on a swamp. A compromise from the

get-go (between Northern and Southern delegates to the Constitutional Convention). Uninteresting topographically, wedged
between two rivers having neither beauty nor utility. It is far
from mountains and ocean. A black city in important respects
that is almost entirely segregated. Surrounded by indescribably
boring suburbs. No federal voting rights, an outrage. There are
no great universities, theater troupes, magazines, art scenes,
bohemian quarters, or ironies. It looks like a suburb with monuments. There are good museums, and that single asset is telling
in itself.

The working city is populated by people interested in one thing
and one thing only: getting an edge politically. Influence. Access.
Power. From the journalists to the NGOs to the lobbyists to the
lawyers and finally to those who are objects of their adoration—
the bureaucrats and politicians—the coin of the realm is politics
and policy. One expects this in capitals of great nations, but usually they intermingle with people not so interested in power and
its privileges. Capitals were often the nation's largest city, attracting all manner of citizens. Washington attracts would-be power
mongers.

And what's wrong with this, other than the poverty of daily
life in the city and its sprawling environs? Political leaders in particular live in a bubble. They like to make fun of "inside the beltway" perspectives, as if that's not precisely what they hold. They
move from insulated neighborhoods in limousines to Capitol
Hill or the Executive Office Building and have nary a moment
with anyone other than their staff, their peers, or favor seekers.
At least if the capital were New York or Chicago or Philadelphia
or Denver, they would occasionally rub elbows with others who
aren't that interested in the continuing resolution or the latest
financial scandal in the Texas delegation or the president's shaky
approval ratings. An occasional conversation like that might be
eye-opening.

The intellectual life is all aimed at the policy process. It's like sleeping on a bad mattress: everything rolls toward the middle. Think tanks are almost interchangeable in their form and even their substance. All is geared to the gatekeepers and other influentials— did you have a senator or representative at your conference, maybe an assistant secretary or ambassador or White House aide or two-star or the foreign editor of NPR or *Congressional Quarterly*? Ah, success. Success is measured by the number of incantations of "our vital national security interests" or "winning the global war on terror." Think tanks are really publicity mills, capturing a "vital" center emblemized by "fellows" who are former assistant secretaries of something who couldn't get the high-rolling jobs on K Street, and now turn to pontificating conventional wisdom in PowerPoint. That is the intellectual class.

The post-9/11 atmosphere in the capital is particularly suffocating. Routes blocked off, Jersey barriers surrounding every building, police presence choking, warning signs ubiquitous, security ridiculously tight. A phony alert from the Department of Homeland Security would mean you wait in line for an hour at the National Gallery of Art for the guards to look through your little girl's pockets. It is nonstop paranoia. I once worked in a small office building on Connecticut Avenue, and every day Dick Cheney's gargantuan motorcade would howl by, replete with motorcycle cops screaming at everyone who looked sideways at the gaudy display. This relentless reminder of the "terrorist threat" has its effects on the average person's mentality, and on politicians' average mentality, too.

So we have a city of intentional insularity in which politics is the only game in town; a city seized by fear of jihadists; a city in which the only legitimate discourse is that of the parties of Tweedledum and Tweedledee; a city in which cultural diversity means a good ribs place in Northwest; and infusing all this mediocrity are the lobbyists and petitioners and journalists and those faux classical monuments all saying, in effect, "This is the greatest power on earth, ever, and you, Senator, are its master."

No wonder the U.S. government acts like Caesar and thinks like Babbitt. Infamously ignorant of the rest of the world, the American people have tolerated the creation of a lobbying haven as its seat of government, a place so drunk with its own grandeur that it can't much be bothered with global inequality or genocides or epidemics—that's for the World Bank, over there on Nineteenth Street. If environments shape the mind, then why aren't we alarmed by the self-gratifying shadow play that is the imperial city?

82   Commercialization of Sports

Pro and amateur sports have always been a part of my life, and I continue to while away many happy hours in arenas of competition. But there is little doubt that most sports have changed in the last fifty years, and the changes are not all good. And many of the bad ones are infecting sports in other parts of the world as well.

The good things are easy to enumerate. Women and girls are now owners of their own expanding space for sports. Television's expansion has brought more sports into homes. Some sports are being taken up seriously in countries where they once were rare, like soccer here.

But the steady drift of nearly all sports toward a much more commercial mind-set is dispiriting and deleterious. College scandals of students being paid off in some way, not graduating, and worse are predictable year after year. Salaries of pro players in the major sports are unimaginably high. The role models of yesteryear were men (almost always) who may have played second base for the Dodgers in the summer and had an insurance business back in

Duluth in the winter. Pro sports were part time. But players stayed with one team and kids and their parents could identify with their team year after year. Those days are gone, too, as are most of the day games, real grass, and seasons with limits.

All of the changes were wrought by money. Television, perhaps, accounts for this as much as anything. Free agency changed the game, too, and with it came a nearly complete surrender to network television and cable and pay-per-view and merchandising and all the other evil influences to generate revenue. On the now-rare occasions when I attend a pro football game, I'm struck by the long empty periods that are, to the television viewer, a stream of commercials. Even on court or field, commercials are everywhere, sometimes on the players' clothing.

The packaging of athletes as hip demigods is now part of the routine. There is a story, possibly apocryphal, that has Michael Jordan and Dennis Rodman practicing one day in Chicago. Rodman, who had just published a "memoir," asked Michael if he'd read it. Michael replied, "Hell, I ain't even read my own."

And the teams and owners have their own anxious dynamic. The superwealthy are always looking for a handout from cities that must feed and clothe their poor. How many episodes of "build me a stadium or we're outta here" can you count?

The saturated marketplace, now replete with junk sports and year-round this or that and sports celebrity worship, is not much about excellence and the joy of playing or watching. It is a marketing industry in which baseball or basketball or football is one of several products, synergystically integrated. The love of the sport, the ideal of team and sportsmanship, has in this business model been outsourced.

This contagion is now spreading elsewhere, by example rather than actual invasion. The breaks for television commercials in soccer matches are an oft-voiced lament. The David Beckham drama in England, when he abandoned Manchester United for Real

Madrid, was daily front-page fodder, and there are many more like it. Juicing athletes, heralded by the steroids flap in this country (but actually pioneered by Soviets and their East European Olympians), is also a disease of commercialization.

Perhaps all of this was inevitable and I'm just a grumpy conservative who wants two baseball leagues of eight teams each with Duke in center, Peewee at short, and Gil at first. But there are more pernicious aspects to this trend than what is conjured by my nostalgia: Is there, at long last, nothing that will not be corrupted by the siren song of money?

83   Damsels in Distress

Was it Chandra Levy? Whoever was the first victim lavished with unwarranted attention, it has become a profitable and distracting habit for the cable news industry. More than annoying, it is asinine and even functions as a political stratagem of sorts.

Chandra was the young woman who went missing while jogging in Washington, D.C.'s, Rock Creek Park in the summer of 2001. A congressman from California had been fooling around with her, and intense speculation about his possible role dominated the summer's news. Meanwhile, Osama bin Laden was issuing warnings and, well, you know the rest. Some mild self-criticism in the news media followed the collapse of the World Trade Center: Chandra was pre-9/11 silliness. It wouldn't happen again.

But of course it does, and with stunning regularity. Now there are entire shows dedicated to missing persons and tawdry little crimes, prime-time purveyors like Nancy Grace or Joe Scarborough. I happened to be in the hospital when the "runaway bride" story came on in May 2005; the man in the bed across the room had the television on nonstop, and I thought the story would last for one or two minutes. *Wrong.* There was Laci Patterson before, and JonBenét Ramsey and Elizabeth Smart and Terri Schiavo and probably a dozen others. But most of them were before 9/11, in the halcyon days of the 1990s when the economy was strong and the nation was at peace. Not much news, it seemed, apart from Bill Clinton's own version of dames at sea.

Then came the poor teenager in Aruba, Natalee Holloway, and that reminded me how 9/11 changed nothing of television "news" habits except for the little American flags that wave on this or that part of the screen.

Like the very light media criticism that accompanied some of these stories, I thought it was just about the white, young, often pretty females that were the center of attention. A few critics pointed out that less white, less attractive, and less female people were disappearing all the time without a wrinkle of notice. In 2001, of all those who were reported missing, women aged twenty-two to twenty-eight were less than 1 percent of the total. The producers of the shows, many of them the morning shows on the networks, will do backflips to justify the "public service" aspect of the coverage, but the more honest in the industry acknowledge that it is what it looks like.

Trivialization of the news is not news. In the opening of *Good Night and Good Luck,* George Clooney's biopic of legendary newsman Edward R. Murrow, the protagonist is delivering a speech in which he blasts the networks' programming as "evidence of decadence, escapism and insulation from the realities of the world in which we live." The criticism of television, as Murrow put it in that same 1958 speech, as "an incompatible combination of show business, advertising and news," echoes through our culture routinely, yet the news now seems driven by entertainment more than ever.

But the Aruba episode, with its thorough absence of news

value and its dominant position on prime-time news shows of CNN and Fox, got me thinking. "Isn't there a war going on?" my wife would lament every time the nonstory would saturate the screen. And there it was: Aruba instead of Iraq. The simplicity of a missing girl and her grieving family, a family "just like yours," clearly and unambiguously the victim of, well, *somebody* in Aruba, including those Aruban officials.

It's an easier story to tell than the complex, contentious, shameful stories in Iraq. Most of those stories are not reported, of course, the tens of thousands of Iraqis dying from American bombs, or the militias executing opponents while U.S. officials turn a blind eye. It's a tough story to report, physically dangerous for journalists and politically dangerous for producers. Afghanistan, even more confusing, by 2005 had only three permanent American correspondents.

Why is this a problem for the rest of the world? There is the sheer distraction from the myriad stories that should inform our understanding of the rest of the planet. There are "good" stories, interesting and tragic, that could grip the viewers if given a chance—the scale of HIV/AIDS, the rapes of girls as a weapon of war, the corruption of tribal leaders from Kurdistan to Katmandu, the bloodcurdling nuclear fantasies of terrorists. These and the more humdrum news of starvation and poverty and the destruction of the world's ecosystems—these are important to our future and, what's more, great tales if told well.

84   Disney, Inc.

One Sunday afternoon a few years ago I took my three-year-old girl to see *Disney on Ice*, an extravaganza of ice skating built around familiar Disney characters and stories. In the finale, all the characters emerged to parade around the rink to the adoring cheers of 12,000 little girls and their parents. I was then struck by the tableau—eight princesses and their savior princes: Cinderella, Snow White, Sleeping Beauty, Belle, and so on, rescued and married to live happily ever after.

I knew many of these stories from my own childhood and saw no particular reason to deny my daughter the same, but I began to notice more carefully how the cartoon characters and stories had changed over time. From the more complex and darker stories of early Disney, including some offbeat characters like the rascally Br'er Fox or the delightful rendering of Robin Hood (featuring Peter Ustinov and Phil Harris), to the present, sharp focus on the same formula of princesses and their dashing boy/men was a long, sad evolution. You know the problem: girls fixating on the princess story, all of them pretty and slender in a certain way, all of them coming of age just in time to be swept off their feet by a young prince who saves them from, typically, something stupid their parents had done.

The marketing for all of these characters is breathtaking—movies, picture books in many formats, computer games, dolls, clothes (for the dolls and for the little girls), stationery, games, and on and on. Barbie, too, is now found as often as a princess as a teenage idol of miniskirts and sports cars. Many classics of children's literature, including *The Little Mermaid* and *Swan Lake,* have been heisted and transformed from charming complexity to the damsel-in-distress mold. Once your little one is hooked, it's an addiction that can last a very long time.

The Disney empire has come in for a lashing from feminists and others concerned about the social lessons embedded in these renditions. The consumer fetishism, racism, paternalism, and adoration of aristocracy have all come in for justified criticism (although the absence of Christian symbolism or proselytizing in Disney stories is admirable). The commodification and homogenization of children's culture is equally onerous. While some of this criticism seems excessive—Disney is hardly the only culprit in this genre—it's an important and troubling set of points in an America where children watch an enormous amount of television and where schools are impoverished as alternative sources of learning.

In the rest of the world, Disney represents something quintessentially American, so these relentlessly Americanized stories, some of which came from other cultures, are an obvious loss for cultural diversity. This form of cultural globalization has many downsides, most tragically the extinction of indigenous cultures. In the United States, Disney theme parks have the additional effect of representing other cultures in a sanitized version that further insulates Americans from real conditions of the developing world.

But Disney has other impacts globally, too. It is one of the worst offenders of sweatshop practices. According to a report of the Hong Kong Christian Industrial Committee, "the stories of exploitation of workers from Haiti, Burma, and Vietnam to Mainland China, producing and supplying to Disney keep surfacing. Long working hours, poverty wage, workplace hazards, awful food, and dangerous and overcrowding dorms are still iron-clad facts revealed in this research on Disney sweatshops." Some of this is just American business as usual overseas, and Disney is focusing, like many of its U.S. rivals, on China not only as a producer of cheap goods for American and European markets, but as a consumer. Overall, its global sales are only $5 billion, but new China ventures—theme parks and television deals—promise much more.

So the wonderful world of Disney is a bit more burdensome than wondrous. In many ways, it is the embodiment of Americanness in both the good and bad meanings—playful, secular, upbeat, but also stultifying, oppressive, and conformist.

85   Las Vegas

Stuck in the Nevada desert, a world unto itself, Las Vegas seems an unlikely choice to make the *100 Ways*. But it does, and here's why: before "Vegas," and all that it conjures up, gambling was a rather low-key affair, confined to a relatively few and often elegant casinos, Thoroughbred racetracks, and back rooms of a million bars where cards and dice ruled the table. It was slightly risqué, and romantic. Think *Casablanca,* Monte Carlo, Saratoga, *Guys and Dolls.* The little guy could get a bet down with the neighborhood bookie, if need be, or take the A train to Aqueduct to see the ponies run. A night at the Casino de Monte Carlo was the stuff movies were made of. Playing poker with your buddies once a month was something men did. Bing Crosby and his pals made Southern California a haven for Thoroughbreds like Seabiscuit and Swaps. "Where the surf meets the turf at old Del Mar, take a plane, take a train, take a car," sang Bing. Even my father, as straitlaced as they came, would get a bet down on the Kentucky Derby and take Mother to Casino Night at the country club.

The innocence of all that is long past, of course. Gambling oozes out of every community pore—lotteries, powerball, slots, video blackjack, Internet gaming, celebrity poker on TV. I recall seeing a poster in a London travel agency window advertising trips to America. The three venues pictured were the Statue of Liberty, Disneyland, and Vegas. Five million come to Vegas from abroad every year. Bad enough that America is being identified as such. But Vegas is not just a destination. Vegas is a state of mind, and an export.

Vegas-style extravaganzas now dot the globe. It is not only the refined casino with "membership" and coat-and-tie mandates that Vegas clones have overwhelmed, but the whole schmear: floor shows, shopping arcades, convention business, multiple gambling emporiums, and all that comes with it. Because of the way it's marketed and presented, gambling goes from naive fun or elegance to something by turns sexually charged, money mad, nonstop, and inevitably tawdry. It does not have to be that way, of course, as there are decades or even centuries of experience in Europe particularly where gambling does not have to engulf and exploit all the emotional weaknesses we possess.

Vegas changed that. It must be a little bit ironic that the desert oasis that was reputedly created by mobster Bugsy Siegel, and grew as an alternative to Havana (Castro shut down the casinos and whorehouses, a blow to some American vacationers and Mafia owners), has become a symbol of American entertainment. Alongside Hollywood, perhaps, Vegas is the most widely known symbol of American entertainment, and now in its many progeny a source of revenue for elementary schools throughout the United States.

It transformed America, with its mysterious allure, sexual energy, and association with glitzy entertainment. It now sets out to take that on the road, to every place that is capable of imitation. It is in Asia, particularly, where Vegas-like cities rise: Asians have money, and many like to roll the dice. Cruise ships and the Internet extend the franchise even farther. (It is a disorienting experience to walk into a hotel in Mozambique or Cairo and see a casino, often called something like "The Strip.") Steadily, it is insinuating itself across the globe.

I can't say precisely why I find this disturbing. Part of it is the sameness of it all. Some of it is the knowledge that casino gambling in particular is not an altogether innocent affair. But Vegasization just leaves a nasty little aftertaste. It's not how I would like America represented, but it is now as much a part of our iconography as the Grand Canyon and the Empire State building.

86   Christmas

The origins of Christmas reach back hundreds of years, a rich and complex history of pagan rituals, early Christian celebrations, festivals of lights at the winter solstice, and many other traditions. Most European countries contributed to this confection of imageries. The Christmas that we know today, however, owes more to Washington Irving than anyone else, who authored the provocative Knickerbocker tales in the early 1800s that shaped many of our conceptions of what the season means—bringing a slightly contentious religious holiday (not celebrated by many Protestants at the time) into the home, while promoting it as a public festivity and virtually creating Santa Claus as we know the jolly old fellow today. Charles Dickens added to the merriment in 1843 with *A Christmas Carol,* embellishing Irving's already Victorian overtones.

As a modern economic engine, Christmas is very much an American invention. Even that most sacred of American myths, Thanksgiving, was moved from the last Thursday of November (established by Abe Lincoln) to the fourth Thursday (by Franklin Roosevelt) to ensure a longer shopping season. But the shopping began long before the 1930s. The rise of industrial production and the middle class in the nineteenth century carved out a niche for both consumerism and a new kind of family that nurtured domesticity, protected children, and adopted American religio-national rituals. These trends and Christmas fit perfectly together, and one might say it was no coincidence that the cultural contributions of Irving and Dickens and others fabricating Christmas legends appeared at the same time. (Some legends, like Rudolph the Red-Nosed Reindeer, were completely commercial inventions; Rudolph was launched by Montgomery Ward in 1939.)

By the 1990s, Christmas was a mainstay of the American economy, representing about one-quarter of all retail sales. The hue and cry denouncing the commercialization of Christmas has been heard for years, but there is little social opprobrium attached to buying presents, decorations, and cards. Americans embrace the secular and material aspects of the holiday season with ever greater gusto. Year-round Christmas stores are now ubiquitous. Christmas buying seasons grow longer each year, it seems, extended by Kmart in 2005 to October 1, for example, and then cascading into January with post-holiday sales. And of course Christian churches have bought into the phenomenon, seeing Christmas as a means of recruitment and donations.

The synergy of Christmas retailing is sometimes breathtaking. I recall a television ad a few years ago in which a father and his preadolescent son were in search of a Christmas tree, repulsed by the crassness of Christmas tree lots. (Tree sales, by the way, topped $520 million in 2003, even more than the $300 million in tree ornaments imported from China.) They drove in their gargantuan SUV, the ad's sponsor, up a rugged landscape to a forested mountaintop, where they came upon a clearing where stood a tall, perfectly shaped evergreen, an exquisite Christmas tree. Dad got out the chain saw and they reverentially approached the tree, but they came to the same conclusion, wordlessly—we can't possibly cut down this natural wonder, it's too precious where it is. So they got back into the SUV to guzzle several more gallons of gasoline whose pollution chokes the forest in search of a tree that someone else had cut down.

One could see in this ad, and hundreds more like it, the sentimentality that now is the core of the Christmas experience. It is not merely the commercialism of the holiday that rankles; it is the substitution of this commodified conformity of sentiments for genuine, yearlong caring and solidarity. Drop a dollar into Santa's bucket and feel charitable. Watch *Miracle on 34th Street,* itself a sophisticated jujitsu of commercial celebration, and get goosebumpy. Call the relatives and send cards to other out-of-touch acquaintances. Attend midnight mass; after all, it's only once a year. Whatever actual normative content Christmas might have—absorbing a lesson or two about the Prince of Peace, for example—long ago was subsumed in the fetishes and habits of the season.

And now it's going global. This Christmas spirit, such as it is, has infected much of the rest of the world. "Christmas in Hong Kong is the time for the tasteless, the season for the syrupy, the holiday for the horrific," writes one blogger of the Chinese metropolis. "For my money, this year's most nauseating display of Hong Kong–style Christmas-hijacking was the Hello Kitty monstrosity infecting the New World Centre in Tsim Sha Tsui." In New Delhi, the capital of Hindu and Muslim India, a major newspaper reported: "Celebrating the festival with enthusiasm, Delhiites seems to have adopted Christmas and established its own traditions. Practically every shop had its own Santa and carol singers singing old favorites, markets, malls, pubs, hotels and even petrol stations were done up in red and green. But commercial establishments keen to cash in on the spirit of buying apart, Santa caps seem to have become a must-have accessory for kids." In Japan, with less than 1 percent of the population being Christian, the season has grown in popularity. Says a tour guide: "Most enthusiastic about Christmas, however, seem to be retail stores and shopping malls, where Christmas trees, Santa Clauses and other seasonal decorations can be found several weeks in advance." In Europe, I have noticed in my visits over the years that what was once a low-key if festive holiday has increasingly followed the American model of a retailing frenzy, replete with the street decorations and blaring Christmas jingles.

Well, I like Christmas, too, and remember fondly the family rituals of my youth. If one has children, Christmas (or some non-Christian variant) is inevitable in America. But it is disheartening to witness the inexorable metastasis of the Americanized version in so many reaches of the globe, not only as an ultimate symbol of cultural hegemony, but as precisely the wrong message of a day that could symbolize an austere and selfless charity but instead extols the virtue and necessity of tacky consumerism. Christmas in the way America has made it is not under attack; it is the predator.

87   The Miami Relatives

Miami has long feasted on its reputation for sunshine, beaches, boating, and carefree image: a little bacchanal oasis in the Old Confederacy. South Beach and its gay and supermodel habitués, "Miami Vice" hipness, Hialeah and Gulfstream high-rollers, sardonic Carl Hiaasen and tart Elmore Leonard, old Miami Beach grandeur, Coral Gables posh—it was all about money and sun and fun. But the core importance of Miami globally is signaled by an altogether different mantra: Kill Fidel, and all his traitor friends. The aquamarine hues of Miami's postcard should actually be blood red: the unofficial capital of Latin America and its violent, hard-core anticommunist right.

The city has been shaped by the Cuban exiles that left when Fidel Castro ousted the corrupt dictator Fulgencio Batista on New Year's Day 1959. From then on, Miami was the center of anticommunist activity for all of Latin America. It was from there that the ill-fated Bay of Pigs invasion, among many other plots to kill or overthrow Castro, was launched.

Groups proliferated, more or less prone to violence, all virulently anti-Castro: the Cuban American National Foundation, Alpha 66, Revolutionary Recovery Insurrection Movement, and Omega 7, among others. It is said that the Cuban American National Foundation, a creation of the Reagan administration headed by Jorge Mas Canosa until his death in 1997, was Miami's most powerful "civic" organization ever, with virtually every wealthy Cuban in the city signed on. It was repeatedly associated with intimidation of moderate or liberal Cuban Americans and others proposing accommodation with Cuba.

The terrorism of some of these groups, often fostered by Reagan or the like-minded in the U.S. government and more local agencies, set the tone for foreign politics in the city. It became a haven for former Latin despots and their henchmen. The city's politics came to be dominated by exiled Cubans and their next generations. The Republican Party grasped the significance of this anticommunism for domestic politics, and Miami—and Florida—became bastions of a 1950s-style politics of hard-edged Cold War fanaticism.

The defining moment of this for American politics came a few months after the notorious Elián Gonzales episode in 2000, when a six-year-old boy, marooned by an escape from Cuba gone bad, was returned to his father in Havana after a six-week national ordeal, punctuated by his vociferous Miami relatives. A few months later, Cuban Americans were storming the vote recount offices of Dade County to preserve the dubious Bush victory in the presidential election.

Whatever its role in domestic and state politics, the Cuban exiles in Miami have created an international city of a special kind, one bestowing legitimacy, not to say ample comforts, on the hemisphere's reactionaries. It is, in the words of Joan Didion, "a tropical capital: long on rumor, short on memory, overbuilt on the chimera of runaway money and referring not to New York or

Boston or Atlanta, but to Caracas and Mexico, to Havana and to Bogotá and to Paris and Madrid. Of American cities Miami since 1959 has connected only to Washington, which is a peculiarity of both places, and increasingly the warp."

It is that strange political geography, unique among American places, that grounds the electromagnetic force with which Miami envelops all of Latin America. Like a beacon of darkness for all the world to fear, Miami and its "relatives" are symbols of the United States's abiding obsessions and permissive political vice.

88   P.R.

It would be presumptuous to claim that America invented public relations. Propaganda and image making are as old as the hills. No one could miss the message in the architecture of ancient Rome, for example, or hucksterism in Victorian England. But it would not be outlandish to say we perfected it.

Edward Bernays, a nephew of Sigmund Freud, is often credited with the invention of modern public relations, and he did indeed bring some insights of sociology and psychology to the manipulation of mass opinion. After opening an office in New York in 1919, Bernays authored numerous works on the topic, including *Crystallizing Public Opinion* (1923), *Propaganda* (1928), and "The Engineering of Consent" in *Annals of the American Academy of Political and Social Science* (1947). Among his early triumphs in a long life of success was the U.S. War Department's Committee on Public Information during the First World War.

What the current War Department does would dazzle even the wily Bernays, however, and here we go from the innocuous—

Bernays's promoting Ivory Soap, and the tens of thousands of commercial P.R. campaigns that followed in his footsteps—to the dangerous. Because what we have now is an institutionally embedded P.R. juggernaut that is manufacturing news and stories and spin at an alarming level of deceit.

Let's just consider a few recent reports:

The U.S. military "is secretly paying Iraqi newspapers to publish stories written by American troops in an effort to burnish the image of the U.S. mission in Iraq. The articles, written by U.S. military 'information operations' troops, are translated into Arabic and placed in Baghdad newspapers" (*Los Angeles Times*, November 30, 2005).

"Public affairs staff at the American-run multinational headquarters in Baghdad have been combined with information operations experts in an organization known as the Information Operations Task Force. The unit's public affairs officers are subservient to the information operations experts, military and defense officials said. The result is a 'fuzzing up' of what's supposed to be a strict division between public affairs, which provides factual information about U.S. military operations, and information operations, which can use propaganda and doctored or false information to influence enemy actions, perceptions and behavior" (Jonathan S. Landay, Knight-Ridder, December 1, 2005).

"The Pentagon awarded three contracts this week, potentially worth up to $300 million over five years, to companies it hopes will inject more creativity into its psychological operations efforts to improve foreign public opinion about the United States, particularly the military" (*Washington Post*, June 11, 2005).

"The Defense Department is considering issuing a secret directive to the American military to conduct covert operations aimed at influencing public opinion and policy makers in friendly and neutral countries, senior Pentagon and administration officials say" (*New York Times*, December 16, 2002).

"Much as the disastrous Bremer regime botched the occupation of Iraq with bad decisions made by its array of administration cronies and relatives (among them Ari Fleischer's brother), so the White House doesn't exactly get the biggest bang for the bucks it shells out to cronies for fake news.

"Until he was unmasked as an administration shill, Armstrong Williams was less known for journalism than for striking a deal to dismiss a messy sexual-harassment suit against him in 1999. When an Army commander had troops sign 500 identical good-news form letters to local newspapers throughout America in 2003, the fraud was so transparent it was almost instantly debunked. The fictional scenarios concocted for Jessica Lynch and Pat Tillman also unraveled quickly, as did last weekend's Pentagon account of 10 marines killed outside Falluja on a 'routine foot patrol,'" (Frank Rich, *New York Times*, Dec. 11, 2005).

One can look back nostalgically to 2002, to Secretary of Defense Rumsfeld's short-lived but memorable Office of Strategic Influence, a Pentagon concoction in which disinformation would be the product to divert and confuse our enemies. When he announced, after a mini-outcry at its disclosure, that the office would be closed, "Mr. Rumsfeld sounded upset," wrote John MacArthur of *Harper's*, "but that was just another aspect of his public relations brilliance. Reading the literal-minded, largely positive reaction to his announcement, I realized that a good many citizens must have inferred that the Pentagon and the White House have been routinely telling the truth over the past few decades."

Every political administration tries to frame the news to help itself, and the recent ones in Washington are no different. But the audacity of the public relations gimmicks does amaze. Think of Colin Powell's phony presentation to the U.N. Security Council on Iraq's WMDs. Or the monikers for corporate giveaways in government programs like "Clear Skies Initiative" and "Healthy

Forests Initiative," or "No Child Left Behind," and one grasps the cynicism about "democratization" and "millennium development goals" and other such global policies that rely on diversionary P.R. rather sensible ideas and actual implementation.

Even Edward Bernays might be a little ashamed.

89   "24/7"

Abuses of the Work Ethic

Work is good. Work that is challenging, important, and rewarding is a blessing. Americans have embraced the work ethic like few nationalities, and the payback is clear in the general prosperity the country has known. But work can also be like a drug, a diversion from other life challenges.

Work has replaced home as the place of comfort, of friends, of personal reward. As Arlie Hochschild so insightfully describes *The Time Bind*, her book about work and home, "Time is being sucked out of homes and pumped into work." But it's not simply that people work more, or must work more to make ends meet. Work is a succor in itself, especially in contrast to the stressful home life that work itself has helped foster. "Most people's friends and even relatives are at work, and if you go home, a lot of people have neighbors but not neighborhoods," Hochschild says. "So when I ask people that I interview—managers, clerical workers, and production workers—where is it you have your friends, you know, where do you talk over problems that you may have at home? At work."

Whatever its causes, work has become an obsession in America, and this is sweeping large swaths of the planet in what can succinctly be termed 24/7: the notion that we're working, or available to work, twenty-four hours a day, seven days a week. To do less is somehow wimpy. We are "No. 1 in the number of hours that full-time workers spend on the job," says historian Stephanie Coontz. "We trail the rest of the industrial world in the number of vacation days we get. And 94 percent of the growth in total income since 1973 has gone to the richest 1 percent of Americans." The work-week has grown gradually, though the number of women in the workforce has gone up sharply over the last two or three decades. For many people, there is no choice: the stagnation in real income or the widespread lack of benefits like health care forces many to work long hours, or more than one job.

But there is also a new ethic among many in the business world, and the vast expansion of worldwide communications—cell phones, BlackBerries, e-mail—enables this as never before. You can never get away, and indeed are never expected to.

Try to find a European in August (and even in July) or around the Christmas holidays and you're out of luck. No one is in before nine. But gradually even this graceful tradition is being eroded by the American pressure to keep up. The saddest sign of this is the slow but certain decline of the siesta.

Economic growth in Spain has meant that the once siesta-proud country is giving up the tradition. Too many connections to the global economy, too much pressure for productivity, too many homes now far from business districts. One report describes "beauty parlors that cater to the sleep-deprived. Silvia Escribano, the owner of a shop called Beauty Boulevard, lets clients catch some Zs in one of the back rooms after a haircut or a back rub. 'It's painful to see them try to stay awake,' she says."

The same has come to Mexico, where late-night hours at work for public employees, driven by long siestas, are now not permit-ted. Two-hour lunches in China are disappearing, too.

Whether or not these changes from a more sensible life-style are driven by American workaholic habits, it is notable that workweeks increasingly conform to U.S. standards. Shorter lunches, fewer lunchtimes at home, working weekends and holidays, and never really being away from the office.

90   Paris Hilton and Celebrity Culture

I have not seen Paris Hilton's television show. I haven't seen her pornographic video. I'm not certain I've ever seen more than a glimpse of her on the entertainment segments on CNN. But I sense, like a spider faintly entering one's peripheral vision, that Paris Hilton is a menace. Or, perhaps more fairly, she may be the apotheosis of a menace—the trash celebrity.

Ms. Hilton is in some ways the perfect foil. She has no discernible talents or achievements. She has a famous name from the world of commerce. She's a willowy bottle blonde of modest good looks, to judge by her press, and limited intelligence. She is part of the Hollywood trend in which many new stars are sons and daughters of other celebrities. (Her grandfather Nicky was one of Liz Taylor's husbands way back when.) Her claim to fame is being famous, a cultivated status that includes party girl misbehavior and over-the-top press attention. The rush of parodies and late-night talk show jokes would take down a lesser mortal, but Paris perseveres. Google her name and it will come up with 30 million links, 28.5 million more than Princess Diana.

Anything wrong with this? Harmless fun? Well, yes and no.

People have always engaged in adoration of celebrities. It's safe to assume that those adored were often men and women of some achievement, even if the achievement was sometimes troublesome—military heroes, for example, or business magnates. And standards change, of course. Christopher Columbus was a hero to schoolchildren of my generation, but was downgraded when his genocidal predilections were publicized some years later. (I've heard that in some private schools that Monday in October is now called "Rethinking Columbus Day," though they still get the day off.)

Hollywood, with its beautiful people and fantasy lifestyles on screen and off, is the perfect breeding farm for celebrity worship. Rock stars, fashion models, and athletes are not far behind nowadays. The Hollywood of yesteryear seems comically tame by today's metrics, with the scandals over Ingrid Bergman or Charlie Chaplin quaint relics of a more innocent age. And the bad boys of sports—remember Billy Martin or Jimmy Pearsall?—pale before the multimillion-dollar megastar stunts of today's arenas.

That the engine of celebrity is huge and hot is not news. The effects of celebrity culture, however, are not wholly appreciated. There are the occasional rants in the more serious precincts of American culture. Lamenting the lavish treatment of Puff Daddy in a *New Yorker* profile, Paul Hollander fumes in *National Review*, "As the *New Yorker* article makes clear, celebrities are handsomely rewarded for the functions they perform. These rewards in turn reinforce their bloated and unrealistic self-conceptions."

The impact of this culture may be more than a simple insult to those yearning for a more elevated expression of thought and art. Celebrity is a form of consumerism, and is deeply embedded in the commercial life of the nation and, increasingly, the world. The wall-to-wall presence of celebrity images is worrisome. Research in psychology suggests the stark influence of celebrity worship in adolescent development. "Intense, personal interest in celebrities was best predicted by low levels of security and closeness," says one study, citing the rite of passage where teenagers emotionally distance themselves from parents and move toward peer groups. Some don't get there.

Such "parasocial" relationships—acting as if one personally knows the celebrity, a neurosis generated by celebrity TV shows and magazines—can be problematic in people of any age group, but seem to affect early teenagers especially, who are searching for identities and role models. Some behaviors, predictable but no less sad, involve problems with self-image, as with strong identification with body types by girls who then develop poor body image for themselves; eating disorders, among other problems, often result.

Another dimension of the parasocial relationships is detachment from the everyday world: "Celebrity worship for intense-personal reasons," concludes a major study, "is associated with poorer mental heath and this relationship can be understood within the dimensions of neuroticism and a coping style that suggests disengagement." And what would that disengagement be from? My guess is that it is not only from the mundane nature of everyday life, but the elements that cultivate sensitivity to, and solidarity with, others. Social relations. Community. Concern for those in the social networks and larger communal issues. One can see the atomization fostered by television greatly reinforced by the focus on celebrities. Parasocial relationships are promoted (to the point of actual addiction) by celebrity television and magazines for commercial reasons, but this also has political consequences.

These studies don't differentiate between the dominant celebrities of today—vapid, self-absorbed, even criminal—and those of an earlier type, who may have represented *something* of social worth. According to *Forbes* magazine's top 100 celebrities of 2005, movie stars and athletes continue to dominate; perhaps there is some hope in that its ranking lists Paris as no more than number fifty-five, just behind Hilary Duff and one ahead of Bill Clinton. (Her ranking, and Bill's, are low due to poor earnings.) But one can take a little comfort in the fact that while athletes and performers work hard to get where they are, not many in that top 100 would be obvious role models for children.

This celebrity mania is projected out into the rest of the world, with trumpets blaring. When visiting southern Africa some years ago, surrounded by post-apartheid turmoil and the looming AIDS crisis, I was asked by a teenage girl in a chance encounter about one of the great icons of America. Bill Gates? No. Oprah? No. She was seeking news about Madonna. Her question in this remote and turbulent corner of the globe was about the Material Girl (who ranks eighth overall on the *Forbes* list).

And around the world it goes. More often than not, it's lamented as being "too American" or too risqué. A glossy magazine editor in India decries "the 'Caucasoid pretensions' among Indians, with fair skin highly prized." A similar dynamic is at work in Africa, where the promotion of a certain image of beauty in African American female celebrities has led to abuse of skin-bleaching creams and changes in perceptions of ideal body types (emphasizing supermodel thinness). "Making it in America," moreover, is associated with the celebrity lifestyles of conspicuous displays of wealth and cavalier attitudes toward everything else.

But perhaps most dangerous is the cultivation of the same kind of celebrity worship that has afflicted Americans. The obsession with the celebrity's life, the pseudo relationships, the detachment from reality—these can be just as isolating in Mombasa or Penang as in Memphis or Des Moines. And isolated individuals are disempowered, emotionally stranded, and fodder for dysfunctional societies and predatory ideologies.

Paris is not forever. But her $300,000 fee "for appearances abroad" and ubiquity on foreign magazine covers signal her global stature as the new U.S. ambassador of celebrityhood. As she told CBS News, "I'm not like anybody else. I'm like an American princess."

91–100   Ten Annoyances

They don't warrant more space, but in some ways it would be a pity to leave these items and people off the list. Some are more important to domestic politics, or culture, or other concerns, while not in the same league as climate change or war or globalization. Some are serious issues and some are not. But they deserve some attention. We could probably generate another *100 Ways* in this vein, but then we'd just be griping.

91   Dumbing Down

Is it possible that the near collapse of primary and secondary education is in fact creating a dumber American populace? It seems improbable, but the evidence is everywhere: two terms for George Bush, reality shows, Nancy Grace, SUVs, Adam Sandler. Test scores in schools are inconclusive (showing slight improvements for nine-year-olds, but none for seventeen-year-olds, which should be alarming); class sizes have increased, and teachers' salaries have not. Very few statistics really send a strong empirical signal about this, but it just feels like the Dumb & Dumber phenomenon is really getting a grip on the country, dude.

Maybe it's the slew of stories about how little our young adults know about world geography. "Last year, a nine-nation survey found that one in five young Americans (eighteen- to twenty-four-year-olds) could not locate the United States on an outline map of the world," says one report from the U.S. Department of Education. "Young Americans knew measurably less geography than Americans twenty-five years of age and over. Only in the United States did eighteen- to twenty-four-year-olds know less than people fifty-five years old and over; in all eight other nations, young adults knew more than the older ones."

Maybe what's most disturbing about this is one (among many) of the findings of this Roper survey: only 13 percent of the Americans knew where Iran or Iraq were, but 34 percent "could determine that the island used for the last season of the television show Survivor is in the South Pacific."

Our ignorance of the world is like a herd of blind elephants on a rampage.

92 Michael Jackson

In the "this is America?" image problem category, Michael must be on the list. Because of his once impressive talent and enormous worldwide popularity, I have to choose him as *the* cultural icon for this category over Tim McVeigh, Ann Coulter, Pat Robertson, and Jenna Jamison. This could go under the celebrity culture critique elsewhere in the book, but he's well beyond the normal dynamics of that.

Michael displays some serious sociopathic behavior. Just find a pictorial history of his face. It's scary. What he does to his body is his business. But what's most annoying about it is how much adoration he still garners from a sizable get-a-life fan base (also true of the others). And, of course, there is the enormous amount of news media coverage, which is truly global in scope.

When I travel abroad and I see the attention in the local media that someone like Michael Jackson gets, I want to get another passport.

93 Slobs

Not often do I consult the Institute of Image Management in Provo, Utah, but I was drawn to their Web comment, "It's Time to Upgrade Our Image America!" Too many slovenly dressed Americans, it said, and so 9/11 . . . Well, not quite. But slovenliness is all too all-American, on this we can agree. And it's getting worse.

My wife and I often fall into a little game when abroad, sitting in a café and people watching. A couple walks by dressed in sweatpants, a football jersey on the male, perhaps, or shorts that are ill-fitting and displaying colors never otherwise seen in nature. "Americans," we'll mutter. We're invariably correct.

We're not alone. "These days, so many Americans dress like slobs at home and also when traveling, wearing their usual blue jeans and 'T' shirt outfits," says International Etiquette Consultant Ruth L. Kerns. "They dress like slobs, and even act like slobs, abroad causing people in other countries to look down their noses at us and to think that we are a bunch of *unwashed barbarians!!*"

This is harsh, but in *100 Ways* we pull no punches. We also seek insights—so what's the underneath to the superficial looks of so many of our compatriots?

I was part of the sixties generation that went to blue jeans and tie-dyes as a statement of cultural defiance. Then, however, there was something to be defiant about (repressive bourgeois conformity, needless to say). Now, with jeans and T-shirts the norm, how to convey withering disrespect for one's elders? Among the young, there are ever more discordant ensembles, the brazen display of undergarments (though didn't that begin with Madonna?) and much body art. The ubiquitous baggy pants on our boys allegedly connote the possession of weapons. But when Sears is selling the pants, you have to wonder.

For the grown-ups, the thirtysomethings or older, casual wear is cheap, hides flab, and conveys at least a fleeting image of youthfulness. It's also a lazy act, like eating fast food. It hardly needs stating that sales of "casual wear" are up, up, up—and as writer

Daniel Akst observes, "the irony is that the more athletic gear we wear, from plum-colored velour track suits to high-tech sneakers, the less athletic we become." It's the sartorial corollary to owning an SUV.

Globally, two aspects of this are troubling. First is the image issue that Ms. Kerns so wisely fingered. Second is the saddening fact that we export our jerseys and baseball caps to the third world especially, so that everyone now can be an Oakland Raider. I saw more University of Michigan shirts in southern Africa than dashikis. At times this spread of pedestrian seediness dips into absurdity. One friend reported seeing a young man in Guatemala with a T-shirt stating "I'm Having a Maalox Moment." He knew not what it meant, one assumes. But then, perhaps in a stroke of subtle irony, he knew exactly.

94 Phalangist Punditry

You know the types—the columnists for daily newspapers and a few magazines that have predictably, tiresomely, and amorally promoted the endless series of wars, support for others' wars and terrorist groups, bottomless military spending, and a "unipolar" attitude of we're the tough kids on the block and now it's time to kick ass. They're the ones who led us into Vietnam, Nicaragua, Iraq, Afghanistan. The results are catastrophic, millions dead, America hated, countries and regions destabilized. Yet here are the armchair quarterbacks yelling from the sidelines that if we don't pound some querulous people into submission we're weaklings. I mean, grow up.

It's an easy-to-spot group—Krauthammer, Podhoretz, Kristol, and a bloating posse of imitators. The neocons tend to be the more obnoxious and less restrained, but whatever they call themselves, they are a contagion. Pompous militarism is the coin of their realm. By weaving an intellectual girdle for the warrior caste in government, they provide support (however superficial) that is reflected throughout much of the news media and the public realm that they otherwise despise. They make shrill noise and render rational discourse impossible.

95 Scientology

The story of Xenu is enough to earn this religion—yes, *religion,* and why not?—a spot on these pages. You see, seventy-five million years ago the ruler of the Galactic Confederacy Xenu brought billions of people to earth (in DC-8s) and blew them up with hydrogen bombs. The souls of these unfortunate, um, people have since been a terrible annoyance to the living. This central narrative of the "space opera" is one of many extraterrestrial contributors to our species, and its problems. Things actually began long before that, you see, 7×10^{85} years to be precise, with Creation Implants . . .

There's more. A lot more. It's not just Tom Cruise and John Travolta being weird. It's a worldwide organization of many divisions and activities, a philosophy of breathtaking lunacy and a very assertive posture. Scientology is all-American, unfortunately, and our major "religious" contribution to the world. And if you think it is no more than a preposterous cult, well, watch out.

96 Hooker Couture for Eight-Year-Olds

Whether we can blame Britney or Christina for this, I don't know, but when I asked my hardly straitlaced friend and mother of two girls what she would put in this book, she said, "Fuck-me clothes for eight-year-olds." And it did not take much time to see she's right. It starts, perhaps, with toys like Bratz Babyz. Consider the maker's ad: "Before the Bratz were everybody's favorite fashion friends, they were the baby girls with a passion for fashion! This is where it all began—the funky fashions, the sizzlin' accessories and the far-out friendship! Look out, these Babyz already know how to flaunt it with lots of Baby Bratitude!"

And flaunting it seems to be the point, even for kids too young to know precisely what they're flaunting. I hope.

"Where there used to be separate fashions and departments for adolescent girls, twelve-year-olds are now being sold the same clothing as eighteen-year-olds," a psychologist told the *Boston Globe*. "It's turning girls into sexualized objects at an earlier age. Who does it serve? It serves the patriarchal culture and the consumer-driven market. As a culture, we're selling sex to girls at a younger and younger age." In advertising especially, but also in the kinds of television programs many kids see, the message of sexual provocation is everywhere. That may be all right for twenty-year-olds, who should know how to deconstruct the message, but not for eight-year-olds.

You needn't be a curmudgeon to see what's wrong here. Growing up too fast, losing childhood, getting into emotional binds—some of them quite serious and durable—all for the sake of a buck. Yes, the demand is there, but it always will be there for the lowest behavior. And, yes, the demand is also percolating abroad, as foreign pop culture often follows America's.

There's a lot to celebrate in America's freedom to challenge social conventions and outdated mores. But that does not mean such challenges are always a good idea.

97 The "Maverick" John McCain

He has that Teflon look. John McCain is the kind of conservative that moderates and even some liberals are reported to love—plain speaking, incorruptible, sensible.

But McCain is a belligerent right-winger who is dangerous on foreign policy and heartless on domestic issues. He may be "good" on campaign finance reform and torture, and he has "character" in standing up to the character assassins taking shots at John Kerry and John Murtha, but these are cost free. On the important stuff, he's with the loony right, 100 percent.

Consider North Korea. He said that the Bush administration had "bungled its North Korea policy through a confused and overly conciliatory approach." Referring to Bush as overly conciliatory is like calling the pope a vamp.

He managed in the first minute of his speech at the 2004 Republican National Convention to conflate all the phony reasons for the war in Iraq—weapons of mass destruction and terrorism. "Only the most deluded of us could doubt the necessity of this war," he said. And then he invoked 9/11.

He resolutely defended Bush's lack of honesty about going to war in Iraq, saying the critics are lying. And he voted to remove Bill Clinton from office in 1999 for lying about, what was it?

He was happy to let Haiti devolve into the nightmare of the Tonton Macoute and opposed Clinton's attempt to restore Aristide, saying we had no responsibility for Haiti. Read a little history, John.

On "intelligent design," the idiotic "creationism" scam, McCain said "let the students decide" whether it should be taught. Let's also ask them if they'd rather read Shakespeare or Spider-Man. He also has predictable right-wing positions on gay marriage, tax cuts, and military spending.

The "maverick" is a menace, essentially a follower of right-wing convention. That's the path to the White House, and that's the path John McCain is treading.

98   Multiculturalism

The best line I ever heard about the urgent fashion in schools to teach "multiculturalism," and the controversies it stirred, was from Todd Gitlin: while we were storming the English department, they were taking over Congress. Multiculturalism—the search for authentic group identities, mainly for ethnic minorities—is a strong force in American education. It has certain attributes that are laudable, especially raising the status of the contributions of blacks, Hispanics, Asians, gays, and so forth in what had too long been a white, European heterosexual culture in America. It is now an issue in Europe and Canada and is gradually spreading across the globe.

In practice, it is a troubling trend. It seems to require separateness and culturally distinct norms. It creates identities for people who may not in fact want those attributes. It is purposefully against any form of universalism and suspects the push for such broad principles as stratagems of the hegemonic group. By insisting on the central importance of difference, it dissolves human solidarity generally, and this dissolution favors domination by the worst aspects of the political culture, which multiculturalism's advocates bristled at in the first place.

By claiming that each culture has its own prerogatives, it is undermining many of the progressive bases of law and justice that have been hard won over the decades, including those of the civil rights movement and feminism. Some cultures deny women a minimum of equal rights, for example, and indeed promote social practices that are extremely harmful to women. Do these deserve protection? By the tenets of multiculturalism, the answer is often yes. The same could be said of child labor, or sexual protection of children, or forced marriages, and the like. If this is an authentic aspect of a certain culture, so goes the argument, then it must be honored.

It is possible, remarkably enough, to have honest accounts of history—such as colonial and racist oppression—and discussions that expose the continuing forms of racial or sexual discrimination in society. Diversity should be honored because it is inclusive and strengthens the whole of society. But too much of multiculturalism is divisive, riddled with guilt-tripping, and dependent on a new form of stereotyping. The struggle for universalism of rights is far from won, and multiculturalism is now, regrettably, one of its adversaries.

99 "Professional" Schools

American universities are the best overall, of this there is little doubt. But an enlarging footprint on even elite universities is apparent—the growth of professional schools, notably business and journalism. Law schools have been around a long time, of course, but represent an additional aspect of the problem.

While one should welcome greater professionalism in management, the news media, and our legal system, it is not a certainty that the explosive growth of the professional school industry is necessary, or wholly desirable. As with all things in America, it affects the rest of the world in example setting, attracting foreigners, and affecting how America acts upon and sees the world.

As is often said about America, there are too many lawyers, and they tend to do what lawyers are trained to do—make laws and regulations that only they can interpret. Ten new law schools are projected to open soon. As it is, there are nearly 200 accredited law schools, a number of state-sanctioned schools, and 735,000 lawyers working in the United States. A number of studies have been done suggesting the quality of lawyer training is inadequate, and the ethics of the profession—particularly the commitment to public service—is a constant source of complaint.

The MBA business is also booming. There were 3,000 MBAs awarded in 1956. Now there are more than 100,000 per year. They are now growing abroad apace. Executive education at these schools is an enormous money maker—up to about $4 billion annually. But there is sharp criticism of what these schools are producing. Irrelevance and lack of problem solving skills are among the brickbats. Says one critic: "'We have built a weird, almost unimaginable design for MBA-level education' that distorts those subjected to it into 'critters with lopsided brains, icy hearts, and shrunken souls.'"

And journalism schools—what is it exactly that they teach? What is it that one can learn that will improve skills as an investigator or writer that other disciplines cannot provide? Journalism schools do not and cannot teach problem-solving skills—critical thinking—as well as the social sciences or humanities or natural sciences. "I'd rather hire somebody who wrote a brilliant senior thesis on Chaucer than a J-school M.A. who's mastered the art of computer-assisted reporting," writes one longtime editor. More than half of the new hires in newsrooms hold *undergraduate* degrees in journalism. There is an ethic among many in this field of "happy amateurism"—that one can report on anything using certain standard methods. You need not *know* anything about the covered topic, which means you'll be at the mercy of people purporting to know—usually the well-heeled who can afford fancy P.R. operations.

The best among them recognize these problems, but the professional school juggernaut churns out lawyers, MBAs, and journalists with credentials to be "professionals" in their fields but who often can't think through a problem creatively, or ethically, because that's just not what the old school did for them. And as this phenomenon metastasizes throughout the world, we get more of the same.

100   The Minutemen

Is there a more ludicrous or embarrassing (though, admittedly, funny) phenomenon than the geezers chasing down Mexican migrants along the Rio Grande? The Minuteman Civil Defense Corps, as they call themselves, is eight thousand strong and patrolling the borders to keep the illegal aliens and terrorists out of our country! Talk about your get-a-life types. Even Bush had to distance himself from this posse. "We are like two commanders on the same side of the battle," they appealed to the president, "with different ideas on how to close the gates. But the enemy is at the gates, and the gates must be closed now!" "Without intervention, they say illegal immigrants from Mexico will create a subnation in the United States. The director of a grassroots group supporting immigration control and border security said he's a nationalist, a constitutionalist and a parent concerned that illegal immigration is threatening national security, jobs and sovereignty and that the government is doing nothing about it," says their Web site.

Militias in recent American history have a rather dodgy record. The Michigan militia, the ones in Idaho and elsewhere that garnered fifteen-minutes-of-fame attention during the Oklahoma City bombing, seemed to be fading away, but the Minutemen have brought the oh-so-proud tradition of vigilantes back to the front pages. They stand with the Ku Klux Klan, abortion doc killers, the black church bombers, and the legions of others in this peculiar American breed.

They'll fade, too, only to be replaced by another flash-in-the-pan (or was that a flash from their six shooter?) paranoid right-wing conspiracy group. As long as there are guns and cameras and foreigners, they won't go away.

Ten Things America Does Right in the World

After all this ranting, do I need to sing a song of love for my country? Copland, Gershwin, Hemingway, O'Keeffe, Salk, Watson, Chaplin, Chomsky, Fossey, Eleanor and FDR, Eisenhower, Murrow, Reuther, Roth, Owens, Ali, Armstrong, Glenn, Douglas, Dewey, Gates, Dylan, Zinn, Chavez, Burns and Allen . . .

The list is astonishingly long and the people are amazing. The contributions in music and art, in medicine and science, in education and innovation: these among others place America as a premier country of all time. It has been blessed by many advantages that countless historians and commentators have pointed out. It has also used those advantages in often humane, enlightened, and generous ways. It is for these reasons—its virtues, its potential power, its own standards—that we must be vigilant and self-critical when it does not measure up.

America has contributed so many good things to the world that it seems downright churlish not to mention at least ten. But it's not just a random or predictable ten I want to explore. I'm not going to visit the icons of American lore. The ten won't include the great battles that saved Europe from the Nazis or the South from slavery. We won't be going to Kitty Hawk or Harvard Yard or

Edison's lab or the Montgomery jail. Too often history is written as if random characters, by pluck and luck, made good and possibly did good. And sometimes, as with Thomas Edison or Martin Luther King, it is easy to dwell on the extraordinary characters and neglect to tell the whole story—what happened, or didn't happen, after their genius dimmed.

I prefer to concentrate on principles and ways of being in the world.

From the Declaration of Independence to the outpouring of generosity for tsunami victims, Americans have embraced values of equality, self-governance, fairness, and mutual aid. That people do this more consistently and surely, with a firmer sense of purpose, than do institutions of government or business is not surprising. After all, most people try to live by some principles. Institutions do this most often when they are obligated to. And if America as a whole—its people, society, government, and businesses—do not always attain the virtues our self-image proclaims, it sometimes does, and that is quite important in the world. It is worth celebrating and encouraging.

Let's explore what those ten might be.

Fairness

The simple sense of fair play for all is embedded in our national culture and much of our law. The American Revolution itself was sparked by a sense of unfairness in the treatment of the colonies by the English king. From the earliest years of the Republic, there were strong social, political, and economic values and institutions that reflected this sense of fairness—public markets to sell food and other goods, for example, which were places of social as well as economic commerce, regulated by the city to sustain fairness, cleanliness, and comity. The timeless popularity of works like *To Kill a Mockingbird* ("One time Atticus said . . .

you never really knew a man until you stood in his shoes and walked around in them") or *The Grapes of Wrath* are fundamental to our self-image. Fairness resonates strongly in most societies (and, scientists assure us, in other primates too), but few have raised it to an icon of nationhood. Fair play. Fairness doctrine. Fair trial. Firm but fair.

Today, "fairness" is a handy political value for use by everyone. But most people can spot the real thing when they see it—fairness for the workers whose pensions were taken away by corporate machinations, fairness for the children of illegal immigrants, fairness for the old, fairness for the disabled, fairness for those who work hard and play by the rules. Perhaps more than any other sentiment, too, fairness is the root of human rights.

The sense of fairness, more than any other, is the underlying principle of opportunity. That everyone has, or should have, a chance to succeed, an education, a job, a home, a dream. To deny anyone those opportunities capriciously is unfair. It is the powerful idea of equality, which was just entering political discourse at the time of the Declaration of Independence, which braces this dedication to opportunity. Equality is, in this sense, the simple commitment to fairness in treatment, not in outcomes, but it is also a principle of law, our guiding ethics, and our daily practice. *Dedicated to the proposition that all men are created equal.* We can disagree over what that means precisely, but it is undeniably a fundament of our political and legal system, and our social practice. And the United States was among the first to embrace that set of ideas.

Our sense of fairness as a deeply rooted value is not only an exemplar for the world, but is indispensable to our moral position as an actor *in* the world. As Dwight Eisenhower said, "Though force can protect in emergency, only justice, fairness, consideration and cooperation can finally lead men to the dawn of eternal peace." Here, here.

The Open Door

It may be that the single greatest thing about America that distinguishes it from almost all other countries over many decades is our welcoming of migrants. We are a nation of immigrants, with an increasingly diverse influx. Even now, after the attacks of September 11, 2001, we are open to large numbers of people who seek a better life. Each year in the 1990s and this decade, we took nearly one million legal immigrants. (In the 1880s, we were taking 500,000 annually.) Nearly 300,000 more are not legal, some entering illegally and others coming in on temporary visas and staying. Not much is made of this, even now (except for occasional spikes of political expediency), because they perform important jobs, contribute to their communities in numerous ways, and gradually work their way into the commons of American life.

Debates swirl about the costs and benefits of migration, the burdens on or privileges of illegal immigrants, and the security issues now attending migration, but the general consensus among the knowledgeable is that the net economic costs are minimal, and may even produce a net benefit. The enriching of culture and the moral obligation of sustaining our ethic of openness trump any crankiness about the porous borders.

The interesting places in the United States are almost always those with many new immigrants. Consider why: those who come are the courageous and diligent. They come for the economic opportunity, and because people are treated better here than in their country of origin. Those are clear incentives to work hard and honor the rule of law, which they do.

Immigration is always a tendentious issue for Americans, in part because it's one of those fostered distractions that can keep us divided. For those who know the whole immigration story, and have frequent interaction with recent immigrants, there really is no controversy. And that's as it should be, as America's greatest legacy to the world.

Honoring Diversity

Immigration has bred diversity, but the freedom to be what you want to be in America goes beyond your national origins. Diversity can mean many things. Lifestyle choices, language preferences, holding fast to the old ways, experimenting with new ways, living in enclaves, integrating anonymously, living a sexual style, embracing the inner you. As one wise man said, it is "possible for me to do one thing today and another tomorrow, to hunt in the morning, fish in the afternoon, rear cattle in the evening, criticize after dinner, just as I have in mind, without ever becoming a hunter, fisherman, cowherd, or critic." We honor, uneasily at times, the many cultures and subcultures that have grown in our verdant gardens. That some people want more conformity to a contrived "American way of life" is ever present, and always a loser. As in nature itself, diversity and evolution always win.

This triumph of diversity is the greatest legacy of the sixties generation, when the black power movement and the white counterculture overturned all assumptions about social life and cleared paths for feminism, gay activism, a blossoming of music and art, and the perpetual demise of bland conformity. Of course it began earlier, in the Beat Generation, in the Lost Generation, in women's movements of a hundred years ago and the cycles of social experimentation that persisted in radical niches of America. But the sudden cultural explosion of the sixties made it a mass phenomenon. Diversity was the engine and the consequence.

As one friend put it to me, America's embrace of diversity is humanity's proof that regardless of how culturally different we are, if given the chance we can live together in a properly functioning society. This is a powerful symbol of hope.

And that is why this emphasis on diversity is important for the rest of the world. It's a symbol and a model. If conservative America can accept and even celebrate gay culture, marriage, equality, freedom, and the like, then other societies surely can as well. If America can readily absorb and honor its many religions and languages and ethnicities, then other places are held up to this standard. It is one of our strongest moral suits, properly so and worth nourishing every day.

Creativity

One could easily make a case for America being the most creative society on earth. Scientific brilliance has long been a hallmark of the country, with dozens of Nobel laureates to show for it. Since the Second World War, artistic brilliance has grown alongside. Pop music worldwide is mainly now American in origin—pop, rock, blues, jazz, and even some classical. Name the top ten painters of the postwar era and they could include Rothko, Rauschenberg, Warhol, Pollock, de Kooning, Wyeth, Shahn. Film directors? Hitchcock, Huston, Scorsese, Allen, Coppola, Lee, Spielberg. Writers would include Mailer, Angelou, Roth, Didion, DeLillo, Baldwin, Miller, and Williams. Yes there are other cultures and other ways of measuring or appreciating creativity; this is not a contest. It is simply to salute the remarkable variety and durability of the American creative genius.

Why does American society yield so much creativity? Its size and wealth certainly account for some of it. But other qualities and conditions come to mind—freedom of expression perhaps above all, the melding and confrontation of cultures, a vibrant civil society that supports its artists, a dense urban experience, and who knows what else.

However one accounts for this flowering, America is an astonishingly dynamic place. Creativity both fuels that dynamism and follows from it. Long may it wave.

Educational Excellence

What grows from our great universities and our once-great public school system are, among other things, the astonishing science, technology, arts, and broad erudition that improves and enriches our lives. I recall vividly, growing up in Indiana, the pride that people had in the great land-grant colleges of the Midwest, and indeed they were excellent universities. One could go to Indiana University, as I did, and pay no more than $300 tuition and receive a first-rate education. Among those educational benefits were introductions to the rest of the world in some important ways.

The universities remain outstanding today, and there is little doubt that the top fifteen research universities in the United States are superior to all but a very small number of universities in the rest of the world. That the primary and secondary school systems have declined in urban and many rural areas is a great and perplexing pity. But educational excellence, and the somewhat equal access to education, remains as America's most powerful asset. And of course these universities and their open doors attract hundreds of thousands of foreign students annually, literally educating the world.

It is commonplace now to say that we live in a knowledge society, that economic growth and global connections are enabled by innovation. The engines of American greatness are the schools, the libraries, and the other tools of pedagogy, critical thinking, and learning.

Secularism

"Congress shall make no law respecting an establishment of religion, or prohibiting the free exercise thereof." There are many countries where the legal system does not contain anything like the establishment clause in our Constitution. While Americans profess to be among the most religious in the world, society and state are secular, and thank heaven for that. No society with strong religious strictures is progressive. No nation in which people are made to feel that they must worship is worth living in. That is because religious belief, particularly the imperative to godliness (such as public displays of religiosity) and the social sanctioning of ungodliness, is relentlessly dogmatic and backward, and all religiously based societies have *always* been so.

All the great institutions of American society are essentially secular—the universities, art and literature, music, athletics, science, medicine, the military, news media, and business enterprise. Each may have religious people or clubs or other associations, but their organization, goals, norms, and operations are free from any particular metaphysical doctrine, including religion. That is one reason why so many of them function so well.

The beauty of it, of course, is that people can believe and practice their faith if they wish, can have communities of faith and institutions in which everything pivots on belief. So these different ways of living can live side by side and even cohabitate. But the fundamentally secular nature of the government, society, and institutions remain independent and therefore strong, vibrant, diverse, creative, open, and powerful.

Aiding the Weary

We have a generous streak, and while it is not always as charitable as it might be—especially when politicized in some fashion—the people of America do give when the need is clear. In response to the Asian tsunami and Hurricane Katrina, there was an enormous outpouring of donations that would put our own government to shame. Each year, the American people donate in cash about $250 billion, averaging about $1,000 per person nationwide. And of course the civic traditions of volunteerism and the growing civil society organizations suggest ample time is donated as well.

Remarkably, much of this comes—as a percentage of income—from the poor. (Mississippi, Arkansas, and South Dakota are the most generous states despite their poverty.) One has to believe that if there were a clearer understanding of the terms and size of foreign assistance, in which the United States is the largest donor in dollars but low in percent of GNP, the people would press for more giving from those accounts as well.

What all the giving benefits is somewhat mysterious. Are donations to churches—the largest single recipient—money spent on the poor? It's not easy to answer that. By its nature, private philanthropy can be difficult to track. And there's some worry that the historic rate of giving—2 percent of personal income—is showing signs of fatigue. But with the boomer generation coming into its twilight years and the enormous transfer of wealth from their parents under way, the giving should remain fairly high. We can only hope that it not only remains generous, but is well spent.

Citizenship

While much of this book has lamented the ways in which social solidarity has been dissolved in recent years, the ideal of citizenship—the ordinary person as the equal to anyone, and one who is empowered to take political action—remains viable, and a keystone of our place in the world.

Citizenship depends on a social, rather than an individualistic, ethos. It recognizes that we are part of the larger whole and that safeguarding the benefits of the polity and the community is our own participation in the life of the neighborhood, the city, and the nation. And participation is a social as well as a political act.

In some respects, the sphere of civic action has increased. Since the Second World War, the numbers of NGOs, including many that are dedicated to protecting and expanding rights, promoting peace, and working to sustain and renew the ecosystem, among other worthy projects, have expanded dramatically. This civic activism has spread to much of the rest of the world, and the models of organizing, action, and indeed the philosophy of civil society—progressive, optimistic, transparent, problem solving— has been at the core of velvet revolutions and transformations since 1989.

The broad availability of citizenship, the commonwealth of citizenry, and the tradition of civic action are the core substance of democracy and the protection of rights, and they are as American as apple pie.

Rule of Law

The old adage "a nation of laws, not of men," is essentially true of America. The laws are not always to our liking, and are not always made for the best of reasons. Far from it. But the respect for this principle, and its utility, should not be underestimated.

It comes from a distant past in which kings ruled capriciously and the law was bent to their liking. The struggle for the rule of law has been arduous over the centuries, and always is contested by tyrants of one kind or another. The laws of elected legislatures, implemented faithfully by executives, and ruled on impartially by judges is the ideal, and America has set that standard as much as any other country, perhaps more so.

While it's easy to deconstruct this notion, those who have lived in places where the law meant nothing would tell a different tale. Laws and the legal system generally favor those with the knowledge and other resources to use it, and that is a sobering fact. But action to advance progressive causes need not be obstructed by the law, and—as the civil rights movement did—can use the law to great effect.

Human Rights

While this also has a contestable pedigree, the United States has long stood for protection of political rights. That this is limited (excluding many economic rights, for example) and not consistently upheld is not to diminish its essential and galvanizing importance in world history.

America's revolution and the rights embodied in the foundation documents were exceptionally important to the development of human rights throughout the world. There were of course other important advances before and after, and in other places, but few that rivaled the exemplary progress between 1776 and 1789 (from the Declaration to the Constitution) that fostered representative, constitutional government and the political rights of freedom of speech, assembly, and so on. Later steps forward—the end of slavery, broadening the voting franchise, recognizing equal rights of women, and establishing a minimum of economic security for citizens—broadened and deepened the initial commitment.

The human rights revolution that changed the world in the 1960s onward was significantly a project of American civil society, including organizations like Human Rights Watch and diverse movements solidifying the rights of blacks, women, and gays. In fact, this gradual broadening of rights was always a struggle with a government that, once established, resisted the enlargement of the rights idea. From the abolition movement that began early in the history of the Republic, to the movements for labor rights, health and safety rights, educational opportunity, and a host of others, the human rights that now radiate from America across the world were hard won by citizen activists. A broadening rights agenda eventually attracted a number of conservative causes, such as demanding freedom of religion in China. It is of more than incidental importance, too, that the military actions in Afghanistan and Iraq were officially justified by appeals to this rights revolution. What the human rights community faces now is not a problem of acceptance, but of continuously nourishing the idea and seeing to it that they are made real in everyday life the world over.

So there—a strong ten things to be proud of, big ideas and historic contributions of the American experiment. These are the values to build on to avert the kinds of mistakes chronicled earlier. I do believe there's a big heart, as well as a will to do good, in the American people. Acting on principle and the best traditions of American history will lead us to a better and more cooperative relationship with the rest of the world.